THE STARS:
THEIR STRUCTURE AND EVOLUTION

THE WYKEHAM SCIENCE SERIES

for schools and universities

General Editors:

To broaden the outlook of the senior grammar school pupil and to introduce the undergraduate to the present state of science as a university study is the aim of the Wykeham Science Series. Each book seeks to reinforce this link between school and university levels, and the main author, a university teacher distinguished in the field, is assisted by an experienced sixth-form schoolmaster.

THE STARS: THEIR STRUCTURE AND EVOLUTION

R. J. Tayler – University of Sussex

WYKEHAM PUBLICATIONS (LONDON) LTD
(A subsidiary of Taylor & Francis Ltd)
LONDON AND WINCHESTER
1970

First published 1970 by Wykeham Publications (London) Ltd.

Cover illustration—Photograph from the Mount Wilson and Palomar Observatories—NGC 5272. Globular star cluster in Canes Venatici. Messier 3. Enl: 1·7×. 200-in. Hale. Hertzsprung-Russell diagram superimposed.

ISBN 0 85109 110 5

Printed in Great Britain by Taylor & Francis, Ltd.
10–14 Macklin Street, London, W.C.2

Distribution:

UNITED KINGDOM, EUROPE, MIDDLE EAST AND AFRICA
Chapman & Hall Ltd. (a member of Associated Book Publishers Ltd.), 11 New Fetter Lane, London, E.C.4 and North Way, Andover, Hampshire.

UNITED STATES OF AMERICA, CANADA AND MEXICO
Springer-Verlag New York Inc., 175 Fifth Avenue, New York, New York 10010.

AUSTRALIA AND NEW GUINEA
Hicks Smith & Sons Pty. Ltd., 301 Kent Street, Sydney, N.S.W. 2000.

NEW ZEALAND AND FIJI
Hicks Smith & Sons Ltd., 238 Wakefield Street, Wellington.

ALL OTHER TERRITORIES
Taylor & Francis Ltd., 10–14 Macklin Street, London, W.C.2.

PREFACE

MANY branches of physics such as gravitation, thermodynamics, atomic physics and nuclear physics are combined in determining the structure of stars. Physical conditions in stars are more extreme than on Earth and a successful understanding of their structure should show how valid it is to extrapolate established physical laws to these conditions.

Considerable progress *has* been made in explaining the observed properties of stars, but many observations are not fully understood. A major aim of this book is to introduce the reader to work in a developing subject and the uncertainties of present theories are often emphasized.

Apart from the use of some special units, such as parsec and electron volt, all numerical quantities are expressed in SI units. These are explained, for example, in the Royal Society booklet *Symbols, Signs and Abbreviations* (1969) and the abbreviation of units and the labelling of graph axes should be especially noted. A list of the more important symbols used in the book and of numerical values of physical constants to the accuracy needed is given on pages ix and xi. As this is probably the first astronomy book using SI units, readers must expect to meet c.g.s. units in their further reading.

Many workers have contributed to our present knowledge of stellar evolution but there are few names in the text, as it is impossible to apportion credit for every advance in the subject in a book of this size. Most of my diagrams are based on the results of other astronomers and my debt to them should be apparent. I am grateful to Mr. D. H. Meyer for drawing all of the figures and to Mrs. Pearline Daniels for her careful typing of the manuscript. I am particularly indebted to my schoolmaster collaborator, Mr. Alan Everest, for his many suggestions, which have led to considerable improvements in the book.

R. J. TAYLER

Lewes
January, 1970

To Moya

CONTENTS

SYMBOLS

A	number of nucleons in nucleus
$B_\nu(T)$	Planck function
E	energy
i	angle of inclination of binary orbit
L	luminosity
m	apparent magnitude (p. 13), molecular weight (p. 57), mean particle mass (p. 58), fractional mass (p. 112)
M	absolute magnitude (p. 18), mass (p. 51)
n	number of particles in cubic metre
N	number of neutrons in nucleus
P	period of binary star (p. 22), pressure (p. 50)
Q	nuclear binding energy
r	distance from centre of star
t	time
T	temperature
T_e	effective temperature
u	thermal energy per unit mass
U	total thermal energy of star
U, B, V	photoelectric stellar magnitudes
X, Y, Z	fractional mass in forms of hydrogen, helium and heavier elements respectively
Z	number of protons in nucleus
γ	ratio of specific heats
ε	rate of energy release
κ	opacity
λ	wavelength
μ	mean molecular weight
ν	frequency
ω	angular velocity
Ω	gravitational potential energy

Suffixes c, s and $\odot$ refer to values at the centre and surface of a star and to solar values respectively.

NUMERICAL VALUES

Fundamental Physical Constants

a	radiation density constant	7.55×10^{-6} J m^3 K^{-4}
c	velocity of light	3.00×10^8 m s^{-1}
G	gravitational constant	6.67×10^{-11} N m^2 kg^{-2}
h	Planck's constant	6.62×10^{-34} J s
k	Boltzmann's constant	1.38×10^{-23} J K^{-1}
m_e	mass of electron	9.11×10^{-31} kg
m_H	mass of hydrogen atom	1.67×10^{-27} kg
N_A	Avogadro's number	6.02×10^{23} mol^{-1}
σ	Stefan Boltzmann constant	5.67×10^{-8} W m^{-2} K^{-4}
$\mathcal{R}$	gas constant (k/m_H)	8.30×10^3 J K^{-1} kg^{-1}

Astronomical Quantities

$L_\odot$	luminosity of Sun	3.90×10^{26} W
$M_\odot$	mass of Sun	1.99×10^{30} kg
$r_\odot$	radius of Sun	6.96×10^8 m
$T_{e\odot}$	effective temperature of Sun	5780 K
parsec (unit of distance)		3.09×10^{16} m

CHAPTER 1

introduction

THIS book is concerned with the structure and evolution of the stars, that is the life history of the stars. Its aim is to show how observations of the properties of stars and knowledge from many branches of physics have been combined, with the aid of the necessary mathematical techniques, to give us what we believe is a good understanding of the basis of this subject.

Because the stars are so remote from the Earth it may seem surprising that we can learn anything about their physical dimensions. To hope to be able to describe their internal structure and, still more, their evolution appears extremely optimistic. The mass and radius of a few stars can be measured directly, but for most stars the only source of information is in the light that we receive from them. This gives us some idea about the temperature and chemical composition of the *surface layers* of the star and about the total light output (*luminosity*) of those stars whose distance from the Earth is known. No direct information is obtained about physical conditions in the interiors of the stars, with the possible exception (discussed in Chapters 4 and 6) that the neutrinos emitted in the solar centre may be detected on Earth. The total of observational information which we have about the stars appears a small amount with which to hope to obtain an understanding of their internal structure.

If it seems presumptuous to hope to explain the present structure of stars, it is, perhaps, even worse when evolution is considered, for significant stellar evolution usually requires millions, or even thousands of millions, of years. Thus there are few instances of observation of stellar evolution and what there are can hardly be regarded as simple evolution. Some stars are observed to be losing mass into interstellar space or to be varying in their light output and occasionally a star explodes dramatically as a supernova, but there are no observations of changes in the properties of ordinary stars. For our nearest star, the Sun, there is no need of very detailed arguments to show that significant evolution must be very slow indeed. A small change in the Sun's properties would suffice to make the Earth uninhabitable for man, and man has been on Earth for hundreds of thousands, if not millions, of years. In fact, geologists say that the Earth's crust must have been solid for several thousand million years and that the Sun's luminosity cannot have changed significantly during that time. This gives an idea of the sort of time which is involved when we interest ourselves in the evolution of the Sun. We shall see later that we believe that more massive stars do evolve more rapidly, but even then we are usually concerned with periods of over a million years.

How then, is progress in this subject possible? The main factor is that physics is a *relatively* simple subject with only a small number of fundamental laws. First, in the case of the structure of a star, we must be concerned with the forces which maintain it in equilibrium. At present it is believed that there are only four basic forces in nature (gravitational, electromagnetic, strong nuclear and weak nuclear) and only these can be involved in the structure of stars. The nuclear forces have a very short range and are not effective in holding together large bodies. The overall structure of a star is governed by the attractive force of gravity which pulls the star together and which is resisted by the thermal pressure of the material forming the star.

The main observational fact about stars is that they continuously radiate energy into space. This energy must have been released from some other source and have been transported from its point of release to the stellar surface. Perhaps the simplest idea would be to suppose that stars were created as very hot bodies and have been cooling down gradually ever since, but we shall see in Chapter 3 that it is impossible to reconcile this with the Sun's steady luminosity for such a long time. If this idea is discarded, the energy must have been converted into heat energy from another form inside the star, and it is then necessary to consider whether gravitational energy, chemical energy or nuclear energy might be involved. When the problem of the Sun's energy supply is considered in detail in Chapter 3, it becomes clear that only nuclear energy can meet the requirements and, in fact, essentially only one process, the conversion of hydrogen into helium with release of nuclear binding energy, can do it.

Of course, these facts that seem so obvious today were not always so clear. The structure and evolution of the stars were being studied before the properties of nuclear binding energy were fully understood and it was then thought that there must be a new unknown source of energy such as, perhaps, the *complete* annihilation of matter into radiation. At one time it was thought that the centres of stars were not hot enough for significant nuclear reactions to occur, and at this time Eddington made his famous suggestion that, if the centres of the stars were not hot enough, the nuclear physicists should look for a hotter place. We shall see in Chapter 4 that the development of the quantum theory made this search unnecessary.

Although the basic forces of nature are few, the calculation of the structure of a star is not simple, as there are so many detailed physical processes which must be considered. It is necessary to have expressions for the rates of many nuclear reactions involved in the release of nuclear energy and even the conversion of hydrogen into helium involves several successive reactions. The energy released by the nuclear reactions must be carried from its point of release to the surface where it is radiated. We must therefore discuss whether the energy is carried principally by conduction, convection or radiation and must study the detailed processes involved in this transport of energy. As mentioned earlier, the pressure of the stellar material resists the attractive gravitational force tending to make a star smaller and the thermodynamic state of the stellar material must be studied so as to discover how pressure depends on temperature and density. In the discussion of the origin and transport of radiation and of the pressure of the stellar material, results will depend on the chemical composition of the star. Some information can be obtained about the chemical composition of the outer layers of a star from the occurrence of an element's characteristic spectral lines in the star's radiation, but it must be recognized that this might not be representative of the chemical composition of the star as a whole.

Because he has only limited information about the properties of *actual* stars, the theoretical astrophysicist tends to calculate the structure of a wide range of *possible* stars rather than trying to explain the properties of an individual star. According to the present theoretical ideas, a few basic properties of a star essentially determine its structure and evolution. The most important factors are believed to be mass and chemical composition, and calculations are made for a variety of different values for these. It then proves more useful to ask whether theory predicts a correct relationship between the properties of stars of different mass and chemical composition rather than whether it predicts the properties of an individual star, which are known only approximately. As we shall see in the following paragraph, this procedure has been particularly useful because there are important regularities in the observed properties of stars. The only exception to this treatment of stars statistically rather than individually is that the Sun has received very detailed attention because we have so much information about it.

A major stimulus to the study of the evolution of the stars comes from the fact that, if one studies the values of mass, radius, luminosity and surface temperature for those stars for which values are available, it is found that not all combinations of values of these quantities are equally probable. The radius, luminosity and surface temperature are not independent because the energy radiated by unit area of the surface of a star is essentially determined by how hot it is. If we regard mass, luminosity and surface temperature as three independent quantities we can draw two independent diagrams relating them. It is usual to plot mass against luminosity (fig. 1) and luminosity against surface temperature (fig. 2) and in both of these diagrams most stars lie in quite narrow bands and there are large regions of the diagrams which contain no stars. For example, it is found that on the average the more massive stars are more luminous and have higher surface temperatures than less massive stars. One of the first tasks of stellar structure theory is to try to explain this regularity and it seems possible that there might be a reasonably simple explanation.

It is believed that the three main factors determining the properties of a star are its *mass*, its *chemical composition* (*when it is formed*) and its *age*. Our observations of stars are complicated by their varying distances and the fact that obscuring matter produces an interstellar fog of varying density between us and the stars. Any attempt to interpret the properties of the whole group of stars for which we have good observational

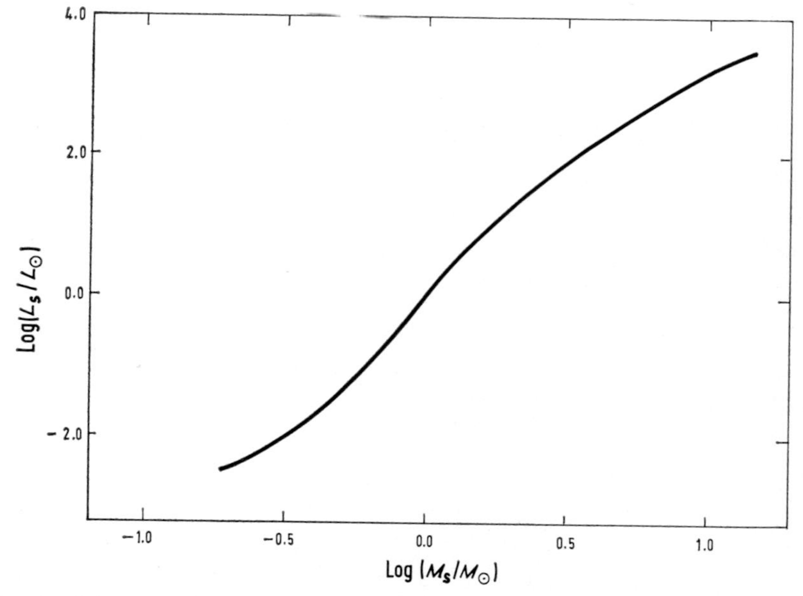

Fig. 1. The mass–luminosity relation. The luminosity L_s is plotted against mass M_s. $L_{\odot}$ and $M_{\odot}$ are the luminosity and mass of the Sun. Stars with accurately known luminosity and mass lie close to the curve shown.

details is also complicated by the fact that the stars vary in mass, chemical composition and age. Such interpretation is easier for the groups of stars which are known as star clusters. These clusters of stars are apparently true physical groupings of stars rather than accidental concentrations, which happen to be in the same direction in the sky, but at very different distances. For a compact cluster it may be hypothesized that, of the five factors mentioned above which contribute to the appearance of a star, four, initial chemical composition, age, distance of the star from the Earth and the obscuring matter in the line of sight, may vary only slightly from star to star. If this is true, *the main factor which accounts for the large differences in the observed properties of the stars is that they have different masses.* This has been the basis of most work on stellar evolution to date and it will be discussed in Chapter 6. Clearly, all of the five quantities do vary from star to star, but it seems reasonable that the variation of mass is most important.

We have already mentioned that the Sun's properties are changing very slowly at present and we further believe that slow variation of observational properties is characteristic of that phase in a star's evolution when nuclear reactions converting hydrogen into helium are

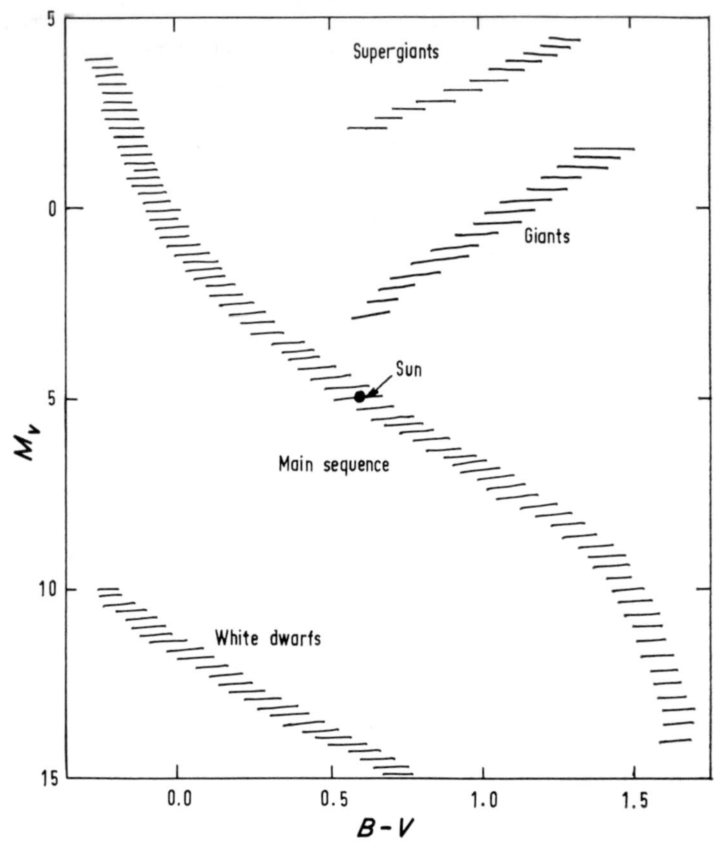

Fig. 2. The Hertzsprung–Russell diagram for nearby stars. The visual magnitude M_V is plotted against colour index B–V and most stars fall in four well-defined groups. (M_V is proportional to $-\log L_s$ and B–V is related to surface temperature, as shown in Table 1, page 17)

occurring in the star's interior and are providing the energy that the star is radiating. This slow evolution prevents us from observing the rate at which stellar properties change, but it also has a very useful consequence in the theory of stellar structure. Because the hydrogen burning phase takes so long, the star settles down in a state which is almost independent of its previous life history. This is useful because even now there is no very good theory of how stars are formed. If the study of stellar structure and evolution depended on having a comprehensive theory of star formation, the subject would have been much slower getting under way. Luckily it has been possible to regard the hydrogen burning phase of stellar evolution as the first stage.

Although the basic physical processes involved in stars, such as those concerned with energy release and energy transport, have been known for 30 years and calculations of stellar structure had already been made even before all of the physical processes were understood, most of the detailed work on stellar evolution has been done in the past ten years. This is largely because, when all of the physics of stellar interiors is to be taken into account, the equations of stellar structure and evolution can only be solved with the aid of a large computer and such computers have not been available for more than about 10 years. The advent of the large fast computer has revolutionized the amount of detail which can be included in the studies. This does not mean that there is no scope for less detailed calculations in which approximate values are used for some of the physical quantities in order to make the equations more tractable. In fact, as we shall see in Chapter 5, the general trends of luminosity and surface temperature as a function of mass can be understood on the basis of simplified physical laws. However, any detailed comparison between theory and observation requires the use of the most accurate mathematical expressions for the physical laws.

It should be stressed that this is a book on a developing subject and that it is not an account of a field in which everything is understood. There are still some serious gaps in our knowledge and it has been my aim to mention and underline these rather than to pretend that they do not exist. Nevertheless we do feel that we have a good general understanding of the subject and believe, perhaps wrongly, that future changes will be ones of detail rather than upsets in the broad principles of the subject.

It is also hoped that what follows in this book will give some idea of how a research scientist approaches a problem. When a subject is completed it may be possible to present its development in a completely logical manner in which each step follows smoothly from the previous one. This is not, however, the situation when a subject is developing. Then it is more like working on a jigsaw puzzle. Pieces must be tried tentatively; the consequences of a variety of assumptions must be tried out. Parts of the subject may be studied in isolation in the hope that, when the whole pattern of the subject emerges, they will fit neatly into it. This is very much the situation with some parts of the subject of this book, particularly the contents of Chapters 7 and 8.

It should be clear from what has been said earlier that the study of stellar structure requires knowledge from many branches of physics such as atomic physics, nuclear physics, thermodynamics and gravitation.

However, it should be stressed that the subject not only makes use of basic physical knowledge, but it also stimulates the development of further knowledge. In particular, we shall mention later that developments in nuclear physics have been stimulated by the need to understand the laws of energy release in stellar interiors and that study of the final stages of evolution of massive stars is leading to interest in the behaviour of the law of gravitation in matter at extremely high densities. It should be remarked that the laws of physics as we understand them, have been obtained from experiments on the Earth and in its immediate environment. In studying the stars and the more distant parts of the universe, *we make the assumption that the laws of physics are unchanging and are the same in all parts of the universe.* This could be an incorrect assumption and, although we always try to understand astrophysical phenomena within the framework of existing physical laws, the possibility that this might be wrong must always be borne in mind.

The remainder of this book is arranged as follows. The observed properties of stars and the techniques of observation are described briefly in Chapter 2. The equations determining the structure of stars are discussed in Chapter 3. Included in these equations are quantities whose values can only be obtained by a rather detailed consideration of the physical state of stellar interiors and the physics of stellar interiors is discussed in Chapter 4. The structure of hydrogen burning stars at the beginning of their evolution, when nuclear reactions have just started to supply the energy the stars are radiating, is considered in Chapter 5. The early evolution of these stars is discussed in Chapter 6, while Chapters 7 and 8 are concerned with later stages of stellar evolution. Finally, some of the problems for future study are described in Chapter 9.

CHAPTER 2

observational properties of stars

Introduction

THE subject matter of this book is stars and in particular the properties of individual stars but, before we start discussing these properties, we give a general description of the Universe in which the stars are situated and of which they may be the most important component. The *may* in this sentence is very important. Until a few years ago there would have been little doubt that stars are the most important constituent in the Universe. More recently it has become clear that there may be a considerable amount of material in the Universe which is not in the form of stars. In giving this brief description of the Universe, no attempt will be made to explain how the results are obtained, but subsequently a detailed discussion will be given of how the properties of stars are deduced from observations.

With the naked eye on a clear night one can observe a few thousand stars and it can be seen that there is a region in the sky, known as the Milky Way, in which there is a particularly large density of faint stars. With even a small telescope, the number of stars which can be seen is greatly increased and it is now known that the solar system belongs to a large flattened system of stars known as *the Galaxy*, which probably contains about 100 000 million stars. Schematic views of the Galaxy as it would look from outside are shown in figs. 3 and 4. The main bulk of the stars in the Galaxy are contained in a highly flattened disk with a central bulge (nucleus), although there are stars in smaller numbers throughout an approximately spherical halo. The diameter of the disk is of the order of 100 000 light years, where a light year is the distance travelled by light in one year† ($9{\cdot}5 \times 10^{15}$ m). The thickness of the disk is only about 1000 light years so that it can be seen that it is very highly flattened indeed.

Many of the stars in the Galaxy are members of binary or multiple systems. A binary system contains two stars which are held together by their mutual gravitational attraction and which describe orbits about their centre of mass. Multiple systems are larger groups of stars which are held together by their mutual gravitational attraction. We shall see below that much of our detailed knowledge about stars is obtained from a careful study of binary systems. Of course, all the

† This unit is not usually used by professional astronomers who use the parsec defined on page 18.

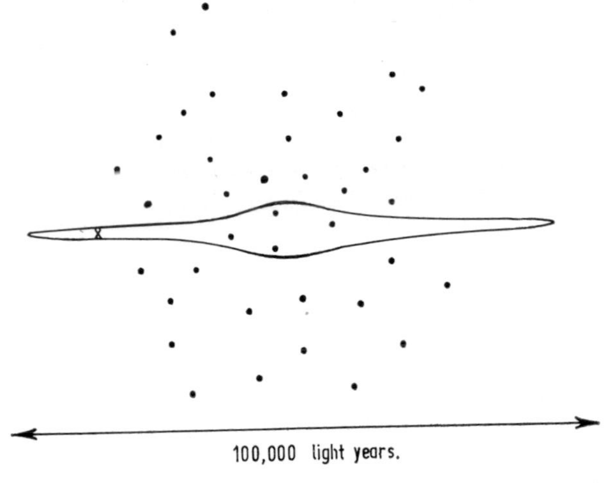

Fig. 3. A schematic view of the Galaxy from the side, showing the thin galactic disk and the central nuclear bulge. The position of the Sun is marked with a cross and the filled circles represent globular clusters.

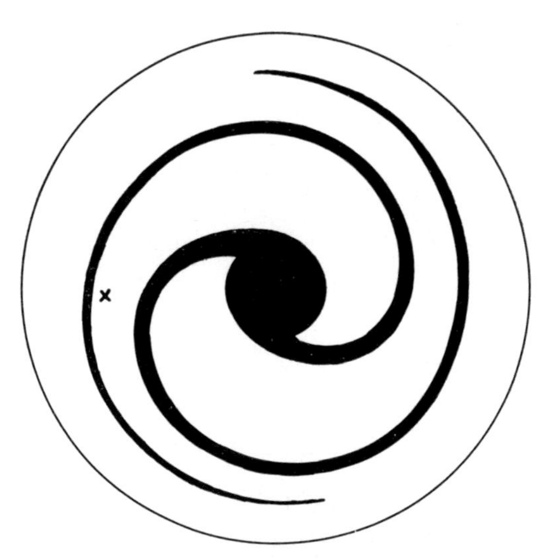

Fig. 4. A schematic view of the Galaxy from above. The spiral structure is shown and the position of the Sun is marked with a cross.

stars in the Galaxy move under the gravitational attraction of all the other stars, but a star which has no very close neighbours moves more or less freely and in a straight line for quite long periods of time. Many stars are also members of larger sub-systems known as star clusters and we shall find later that the existence of these star clusters is very important for the subject matter of this book. Clusters are of two general types, globular and galactic (or open), although there is no completely clear division between the two types. The general appearance of globular and galactic clusters is shown in figs. 5 and 6. Globular clusters have a compact circular appearance, they are spread throughout the Galaxy including the halo, there are at least 100 of them and they

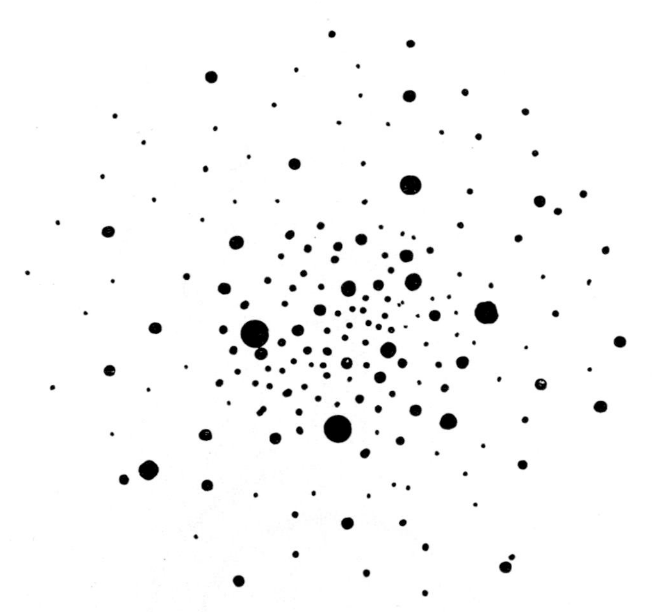

Fig. 5. The distribution of stars in a globular cluster. The brightest stars are shown larger as they would appear in a photograph.

contain between 100 000 and 1 000 000 stars each. The galactic clusters contain many fewer stars. They are called galactic because they are situated in the plane of the Galaxy and open because their appearance is diffuse rather than compact. There are several hundred galactic clusters known.

As shown in fig. 4, the disk of the Galaxy has a spiral structure. Many of the brightest stars in the Galaxy are found in, or near, the spiral arms. As well as stars the Galaxy contains clouds of gas and dust. The gas is also largely situated in the spiral arms and it may account for between 1/20 and 1/10 of the mass of the Galaxy. The interstellar gas is believed to be the material out of which stars are formed. We shall see later that the bright stars in the spiral arms are thought to be stars which have been formed quite recently in the galactic lifetime, and it then seems natural that they are closely associated with the interstellar matter out of which stars are still forming today.

At one time it was thought that the Galaxy was the whole Universe, although there were objects called spiral and elliptical nebulae whose position in the Galaxy was unclear. It is now known that these nebulae are also galaxies and, in particular, some of the spiral nebulae are very similar to our Galaxy. Galaxies have now been observed out to distances of a few thousand million light years and, with galaxies being

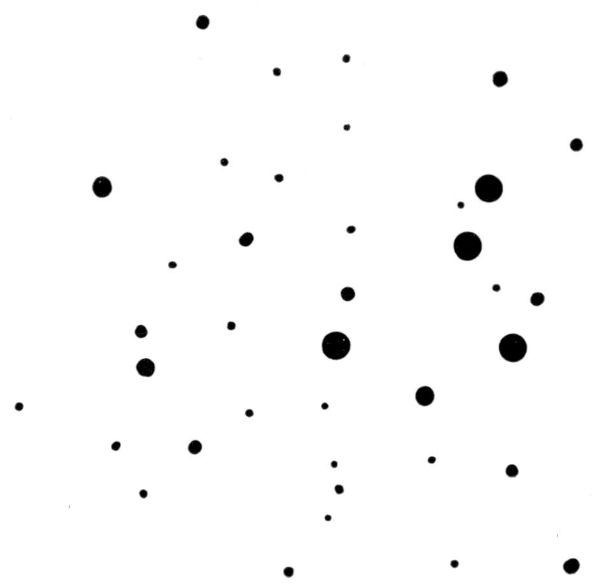

Fig. 6. The distribution of stars in a galactic cluster.

typically a few million light years apart, this means that there are many millions of galaxies. The stars in these galaxies appear to be similar to the stars in our own Galaxy and the theory of stellar structure developed in this book should be applicable to all of them, although it is only in our immediate neighbourhood in our Galaxy that stellar properties can be observed in fine detail. A discussion of the properties of all the galaxies in the Universe would soon involve us in a discussion of cosmological theories. These theories are concerned with whether the Universe had an origin in time or whether it has always existed, whether all the galaxies were formed at about the same time or whether galaxies are still being formed today and many similar questions. We do not discuss cosmology in this book and, in fact, the study of the life histories of stars in our own Galaxy is essentially independent of wider cosmological questions.

We now return to a consideration of the properties of individual stars. We shall find that there are many gaps in our knowledge, but that nevertheless a reasonably consistent picture emerges.

Luminosity, colour, surface temperature

Most information about stars is obtained from the light and the other electromagnetic radiation which they emit. Observations give us some information about both the *quantity* and the *quality* of this light. In principle we can detect the amount of radiation from a star which falls on unit area of the Earth's surface and we can investigate what is the distribution with wavelength of the radiation. There are many different detection systems used. These include direct photography with a photographic plate, which is sensitive to a rather wide wavelength range, and the use of prisms or diffraction gratings to spread the light out into a spectrum before it falls onto the photographic plate. There are also many devices based on the photoelectric effect which detect the electrons emitted from a *light sensitive surface*. In most cases filters are used to cut out all of the light except in a narrow wavelength range and this is known as *narrow band photoelectric photometry*. If the entire energy output of a star, irrespective of its wavelength, is to be measured, a *bolometer* or *pyrometer* may be used; these measure the energy received in the form of heat.

For some purposes it is useful or essential to detect the radiation in narrow wavelength bands, while for others it is more useful to have a detector with as wide a wavelength response as possible. At present the theoretical astrophysicist finds it much easier to predict the total output of radiation from a star of given mass and chemical composition than to calculate its exact distribution over wavelength. Thus for comparison with theory it is desirable to measure the radiation over as wide a range of wavelength as possible either by the use of many narrow wavelength band detectors spread over the whole spectrum, or by the use of a bolometer which responds to the entire wavelength range of interest. In what follows we shall refer particularly to two types of observation. These are spectroscopic observations which are essential for discussions of the chemical composition of stars and photoelectric measurements in three wavelength bands known as U, B and V which are centred in the ultra-violet, blue and yellow regions of the spectrum and which will be defined and discussed further below.

Magnitudes

Observations of the light received from a star are normally expressed in *magnitudes*. Magnitude is a logarithmic measure of luminosity with the brightest stars having the lowest magnitude. This convention arose because the Greek astronomers originally catalogued the stars visible to the naked eye in six magnitudes, with the first magnitude stars being the brightest. When, in the nineteenth century, a quantitative system was first introduced, it was made to agree as closely as possible with the old qualitative measures. Thus

$$m = \text{const} - 2{\cdot}5 \log L, \tag{2.1}$$

where m is the magnitude, L is the luminosity (total light energy received) in the wavelength range being studied and the constant is used to define the zero of the magnitude scale. Such a magnitude scale, with the zero point chosen appropriately, gives reasonable agreement with the earlier estimates because the human eye more closely responds to the logarithm of the luminosity than to the luminosity itself.

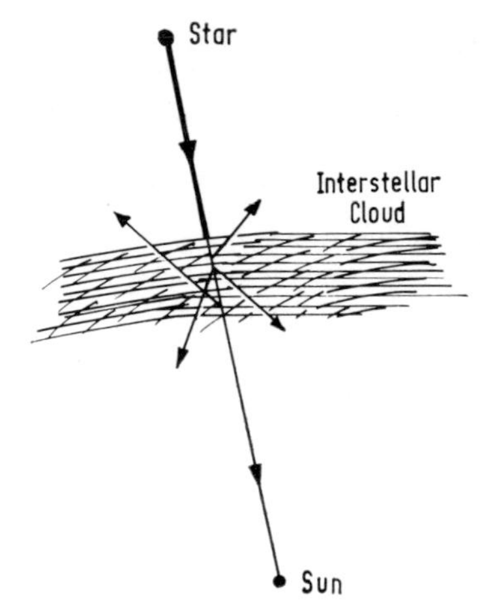

Fig. 7. Scattering and absorption of starlight by an interstellar cloud.

Note that the luminosity entering into equation (2.1) is the apparent luminosity, in the sense that it refers to the amount of radiation which falls on a detector on the Earth's surface. If we are to discuss the intrinsic properties of stars, we must convert this to the absolute luminosity of the star, the energy emitted by the star per second. In order to convert apparent luminosity to absolute luminosity we first need to know the distance of the star from the Earth. Then the amount of radiation falling on unit area of the Earth's surface can be multiplied by $4\pi d_*{}^2$, where d_* is the distance of the star from the Earth. This alone is difficult because there are not many stars whose distance can be measured directly (see the discussion later in this chapter).

However, the problem is more difficult than this because radiation can be scattered or absorbed by material between us and the star (see fig. 7) either by the gas in interstellar space or by the Earth's atmosphere. Until quite recently astronomical observations have been restricted to the wavelength ranges in which there is an atmospheric *window* (see fig. 8). It is at first sight fortunate, but presumably in no way accidental, that the visible *window* almost coincides with the wavelength range to which the human eye is sensitive and that it is a range which contains most of the radiation emitted by the Sun and many other stars. With the advent of rockets and artificial satellites and the possibility of placing telescopes in space above the Earth's atmosphere, this difficulty will be at least partially removed. The same is not true of the effect of interstellar matter between us and stars and, although it is possible to estimate its influence, some uncertainties remain.

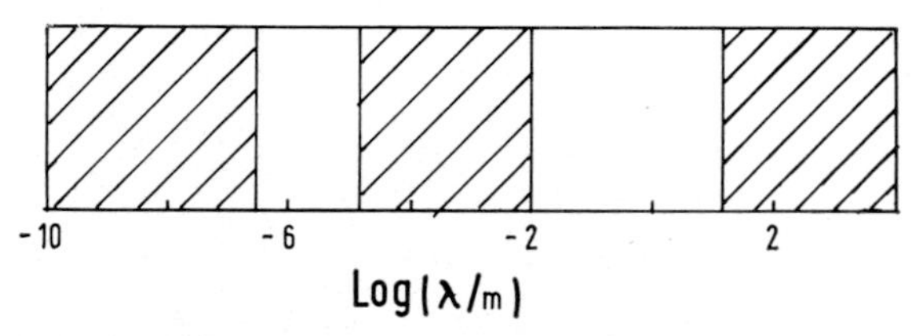

Fig. 8. The transparency of the Earth's atmosphere. Electromagnetic waves of wavelengths corresponding to the hatched areas are almost completely absorbed in the Earth's atmosphere. Between the hatched areas are the visible *window* and the radio *window*.

We can write down an equation which relates the quantity of radiation emitted by the star to the amount observed. Suppose first that $L_\lambda \,\mathrm{d}\lambda$ is the total energy emitted by the star in the wavelength range between λ and $\lambda+\mathrm{d}\lambda$, so that the luminosity (total rate of emission of energy) of the star is:

$$L_\mathrm{s} = \int_0^\infty L_\lambda \,\mathrm{d}\lambda. \tag{2.2}$$

If the Earth's atmosphere and interstellar space were transparent to radiation, the energy reaching unit area of the Earth's surface, per second in wavelength range $\mathrm{d}\lambda$ would be $L_\lambda \mathrm{d}\lambda/4\pi d_*^{\,2}$. We can introduce a quantity t_λ which is the probability that radiation of wavelength λ will reach the Earth's surface and a second quantity s_λ which measures the sensitivity of the detection system used. Then the amount of energy received per unit area at the Earth's surface in wavelength range $\mathrm{d}\lambda$ around wavelength λ is:

$$l_\lambda \mathrm{d}\lambda = L_\lambda \mathrm{d}\lambda\, t_\lambda s_\lambda/4\pi d_*^{\,2}, \tag{2.3}$$

and the total energy received is:

$$l_\mathrm{s} = \int_0^\infty (L_\lambda t_\lambda s_\lambda/4\pi d_*^{\,2})\,\mathrm{d}\lambda. \tag{2.4}$$

Surface temperature

By the *quality* of the radiation emitted by a star is meant its distribution in terms of wavelength or alternatively frequency†. The crudest measure of the quality of the light is its colour and this is often referred to in descriptions of stars as red giants, white dwarfs, blue supergiants, etc. A complete description of the quality of the light involves the measurement of l_λ at all wavelengths. The quality of the light which we receive from a star is not affected directly by the distance of the star from us; the quantity of radiation is reduced by the same geometrical factor, $4\pi d_*^2$, at all wavelengths. The quality of radiation is affected by the Doppler effect if the star being studied is moving towards or away from us. Although this Doppler effect which shifts spectral lines either towards the red or the blue can be used to deduce the velocity of the star, it only has a significant effect on the quality of the radiation if the velocity is comparable with the velocity of light. This is true of some distant galaxies which are receding from us with high velocities, but not of stars in our own and nearby galaxies. However, as is clear from equation (2.3), the quality of the light is affected by absorption and scattering which do not act equally on all wavelengths and this fact helps us to unravel how much absorption has occurred.

The colour of a star is related to its surface temperature, although the latter cannot be uniquely defined. A temperature can be defined when a system is in a state of *thermodynamic equilibrium*‡ and then the distribution of radiation with frequency is uniquely defined by the temperature and follows the *black body* or Planck law. In these circumstances the amount of radiant energy crossing unit area, in a unit solid angle about the direction normal to the area, in unit frequency range and in unit time is $B_\nu(T)$, where

$$B_\nu(T) = \frac{2h\nu^3}{c^2}\frac{1}{\exp(h\nu/kT)-1}. \tag{2.5}$$

In expression (2.5), B_ν is called the Planck distribution at temperature T K, ν is the frequency, c is the velocity of light (3×10^8 m s^{-1}), h is Planck's constant ($6{\cdot}6 \times 10^{-34}$ J s) and k is Boltzmann's constant ($1{\cdot}4 \times 10^{-23}$ J K^{-1}). In practice we often use the word temperature when a state of thermodynamic equilibrium does not exist; in particular we often use it as a measure of the mean kinetic energy of the particles present. For the stars we do not have very direct information about the particles and we must try to deduce a surface temperature from the radiation which we receive. Planck curves for three values of T are shown in fig. 9.

For some stars the distribution of energy with frequency is not too different from the black body curve (2.5) and for these stars a surface

† In the remainder of this book we will describe light by frequency rather than wavelength. The two are, of course, related by $\lambda\nu = c$, where ν is the frequency and c the velocity of light.

‡ The concept of thermodynamic equilibrium is discussed in the Appendix.

temperature can readily be defined. For others this is more difficult and observers now usually use a less subjective measure of the quality of the light known as colour index instead of surface temperature. A colour index is the difference between the magnitude of the star in two wavelength bands. Thus, if we use the three colour U, B, V photometric system mentioned earlier, we can define three colour indices U–B, B–V, U–V, where the symbol U, for example, is now being used to

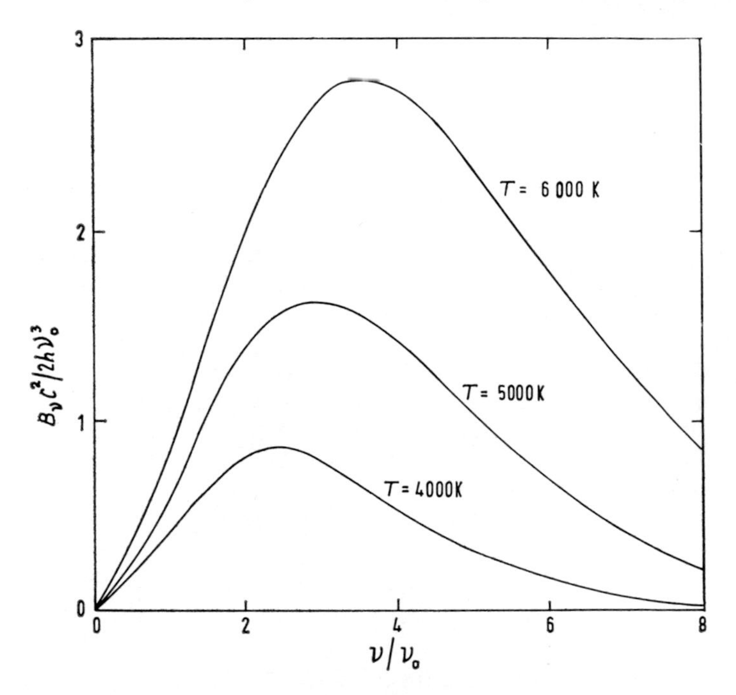

Fig. 9. Planck curves for three values of the temperature. The normalizing frequency ν_0 is 10^{16} s^{-1}.

denote the magnitude of the star in the U band. The filters used for each of these bands admit radiation in a range of wavelength about 1000 Å wide and their central wavelengths are approximately:

$$\lambda_U \approx 3650 \text{ Å}, \quad \lambda_B \approx 4400 \text{ Å}, \quad \lambda_V \approx 5480 \text{ Å}. \tag{2.6}$$

If a star did radiate as a black body the colour index would be directly related to the logarithm of the surface temperature; in general it is an approximate measure of surface temperature which is less subjective than estimated surface temperatures.

Effective temperature and bolometric correction

Later in this book we shall be concerned with making a comparison between observed properties of stars and those predicted by solutions of the equations of stellar structure. As mentioned earlier the theoretical astrophysicist finds it much easier to predict the total amount of energy radiated by a star than to calculate its distribution with frequency. Theoreticians define what they call the effective temperature of a star, T_e. This is defined in such a way that a black body of temperature T_e with the same radius as the star would radiate the same total amount of energy. Thus

$$L_s = \pi acr_s^2 T_e^4 \equiv 4\pi r_s^2 \sigma T_e^4, \tag{2.7}$$

where r_s is the radius of the star, a is the radiation density constant ($7{\cdot}55 \times 10^{-16}$ J m^{-3} K^{-4}) and σ ($\equiv ac/4$) is the Stefan–Boltzmann constant† ($5{\cdot}67 \times 10^{-8}$ W m^{-2} K^{-4}).

$B-V$	M_V	$\log T_e$	M_{Bol}	$B-V$	M_V	$\log T_e$	M_{Bol}
−0·3	−4·4	4·48	−7·6	0·5	3·8	3·80	3·8
−0·2	−1·6	4·27	−3·5	0·6	4·4	3·77	4·3
−0·1	0·1	4·14	−0·8	0·7	5·2	3·74	5·1
0·0	0·8	4·03	0·4	0·8	5·8	3·72	5·6
0·1	1·5	3·97	1·3	0·9	6·2	3·69	5·9
0·2	2·0	3·91	1·9	1·0	6·6	3·65	6·2
0·3	2·3	3·87	2·2	1·1	6·9	3·62	6·4
0·4	2·8	3·84	2·8	1·2	7·3	3·59	6·6

Table 1. The relationship between colour index, $B-V$, absolute visual magnitude, M_V, logarithm of effective temperature and absolute bolometric magnitude, M_{Bol}, for main sequence stars.

One of the main problems in correlating theories and observations of stars is that of relating effective temperature to colour index or other estimates of stellar surface temperature and bolometric luminosity to magnitude measured in some particular wavelength range. Thus, in particular, we are often interested in the transformation from (L_s, T_e) to (V, $B-V$). Ultimately such a transformation can only be made by measuring the total energy output of the star with a bolometer or by measuring magnitudes in a large number of narrow wavelength bands. In practice such observations can only be made for a limited number of stars, but these can be used to set up an empirical relation between effective temperature and colour index and between bolometric magnitude and visual magnitude which can then be applied to other stars. Such transformations for main sequence stars (the term will be defined on page 33) are shown in Table 1.

† There is considerable confusion in the astronomical literature concerning the naming of the two constants a and σ. In particular in many books concerned with the structure of stars, a is called the Stefan–Boltzmann constant while in many other books σ is called the Stefan–Boltzmann constant. Our present usage is now standard.

Absolute magnitude
The definition of magnitude given in equation (2.1) is in terms of the amount of radiation received on unit area at the Earth's surface and it is known as the *apparent magnitude* of the star. We often wish to use a magnitude which is a measure of the total light emitted by the star. The *absolute magnitude* of a star is defined to be the apparent magnitude it would have if it were placed at a distance of 10 parsecs, where the parsec will be defined in the next paragraph. If the actual distance of a star is d parsecs, its absolute magnitude M and apparent magnitude m are related by:

$$M = m - 5 \log (d/10). \tag{2.8}$$

Stellar distance
In order to convert apparent magnitude into absolute magnitude, the distance of the star is required. This can only be obtained directly for a relatively small number of nearby stars. For these it can be measured by a trigonometric method. Suppose we consider the Earth's motion around the Sun (fig. 10). The apparent direction of a star, measured relative to the position of much more distant stars, changes as the Earth describes its orbit around the Sun. If this angular displacement can be measured, the triangle EE′S′ can be solved to find the distance of the Earth (or Sun) from the star. The angle ES′S is called the *parallactic*

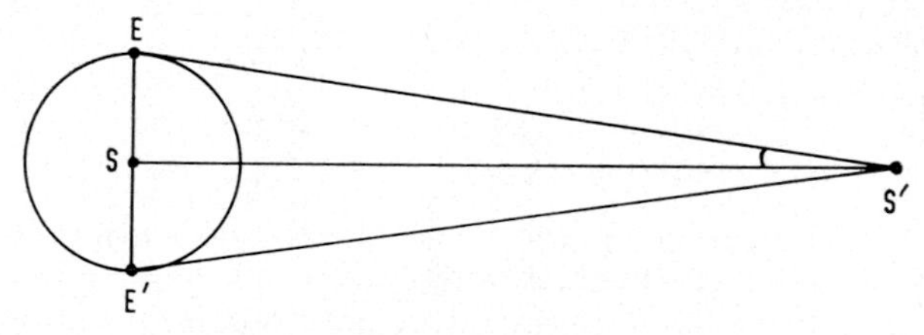

Fig. 10. The parallactic angle of a star.

angle of the star. For the nearest star (other than the Sun) the angle is somewhat less than 1″, which means that the distance to the nearest star is very great indeed. Because of the size of astronomical distances, astronomers use a distance scale based on the distance at which the parallax would be 1″ known as the parsec, where

$$1 \text{ parsec} = 3{\cdot}09 \times 10^{16} \text{ m}, \tag{2.9}$$

or

$$1 \text{ parsec} = 3{\cdot}26 \text{ light years}. \tag{2.10}$$

The discussion of the measurement of stellar distance given above has been rather simplified. Thus it has been assumed that the star being observed is at rest relative to the Sun. In fact the stars in the Galaxy are not at rest, but they are describing orbits in the mutual gravitational fields of all the other stars with velocities relative to one another of the order of 10^4 to 10^6 m s^{-1}. If the star being studied moves in the direction perpendicular to the line SS′ a distance comparable to, or greater than, the distance EE′ in the six months that it takes the Earth to travel from E to E′, as it can with the velocities quoted above, the method described will not give the correct answer (the distance moved in the line of sight is unimportant because even for the nearest stars this is minute compared with the distance SS′). However, the motions of stars are such that, unless they are partners in close binary systems, they can be assumed to move with a uniform velocity in a straight line for periods measured in years, and deviations will only become apparent after thousands or millions of years. This means that, by observations over a period of several years, the steady displacement of the star with respect to very distant stars which is produced by its own motion can be separated from its periodic displacement due to the Earth's motion about the Sun (fig. 11).

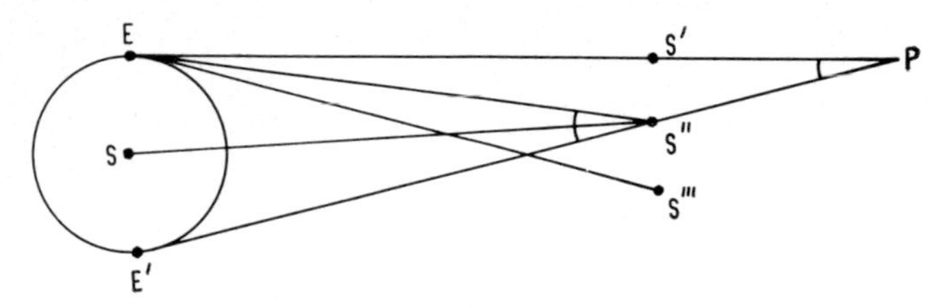

Fig. 11. The parallax of a moving star. As the Earth moves from E to E′ and back to E, the star moves from S′ to S″ and S‴. For a distant star, the true parallactic angle is ES″S which cannot be directly observed. The angles EPE′ and S′ES‴ can be observed and, assuming the star is moving with a constant velocity, the parallax can be obtained by simple geometry.

As the largest parallax for any star is less than 1″, it is perhaps surprising that any parallaxes can be measured. In fact the parallaxes of three stars, 61 Cygni, α Lyrae and α Centauri†, were first measured by three different observers in the year 1838. Parallaxes can be measured with some degree of accuracy down to about 1/50″ (distances of 50 parsecs) and they are known for several thousand stars. In terms of the dimensions of the Galaxy which have been mentioned earlier in the chapter, 50 parsecs is a very small distance and it is clear that direct distance measurements can only be made for the nearest stars. For more distant stars indirect methods of estimating distance must be used and some of these will be mentioned later in this chapter.

† Stars are named by referring them to the constellations in which they occur. The brightest stars are denoted by Greek letters and fainter stars by numbers. Thus α Lyrae (Vega) is the brightest star in the constellation Lyra. The constellations are apparent groupings of stars in the sky in contrast to clusters which are physical groupings.

Proper motion
In measuring distance we have separated out the periodic displacement of a star caused by the Earth's motion from its regular displacement produced by its own motion. This apparent angular motion of stars across the sky perpendicular to the line of sight is called the *proper motion.* Measurement of proper motion gives some information about the way in which stars are moving, but the apparent motion cannot be converted into a true velocity unless the distance of the star is known (see fig. 12). However, observations of proper motions are useful in helping us to discover nearby stars whose distance can be measured. Parallaxes can be measured for nearby stars, but these stars do not carry labels saying *nearby star*. There are two obvious ways of trying to identify nearby stars. The first is to assume that many of the stars which appear very bright are also very near to us. The other is to choose stars with large proper motion. Unless there is a steady increase in velocity of stars with distance from the Sun, which seems unlikely as the Sun is not at the galactic centre, it is likely that stars which have large proper motions are also nearby stars. Thus the first stars to study for parallaxes are apparently bright stars with large proper motions.

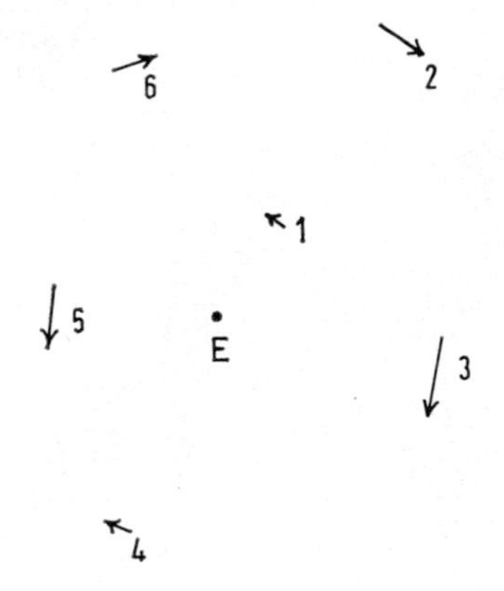

Fig. 12. The motions of six stars in a given period of time are indicated by the lengths of the arrows. Their *proper motion* in that time is the angle subtended by the arrow at the Earth, E. Stars 1 and 2 have different true motions but identical proper motions.

The study of parallaxes, proper motions and radial velocities (velocities in the line of sight deduced from the Doppler effect on spectral lines) gives us information about the positions and motions of stars and hence about the structure of our neighbourhood in the Galaxy. This, however, is not the subject matter of the present book.

Stellar masses
There is only one direct way of obtaining a stellar mass and that is by

studying the dynamics of a binary system. The method used depends on whether the binary system is a wide binary or a close binary. According to Kepler's laws, the two stars in a physical binary system (that is a genuine binary system rather than two stars which are in the same direction from us but which are at totally different distances) revolve about their centre of mass in elliptic orbits (fig. 13). If the stars in a binary are sufficiently far apart, in terms of their distance from the Earth, both stars can be observed and over a sufficiently long period of time their orbits about their centre of mass can be studied. If the parallax of the binary system can be measured so that its distance from the Earth is known, the apparent size of the orbits can be converted to a true size. The size of the orbit combined with the period of revolution of the stars enables the masses of both stars to be determined by use of Kepler's laws. This is discussed on page 22.

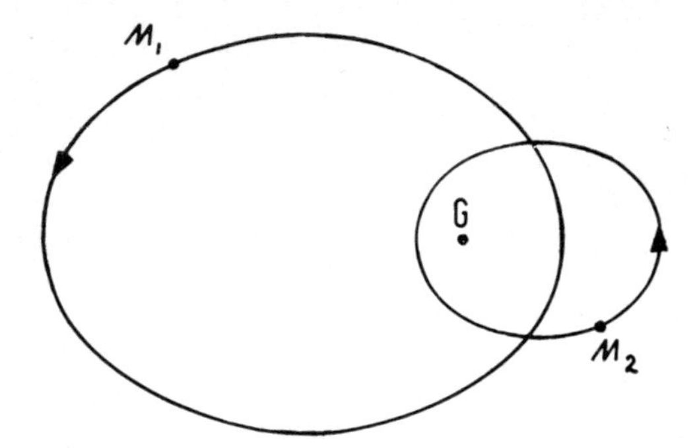

Fig. 13. The elliptical orbits of two stars of masses M_1 and M_2 about their centre of mass G.

In fact the problem is not quite as simple as that. In fig. 13 the orbits of the stars have been drawn as they would appear if the orbital plane were perpendicular to the line of sight from the Earth. If sufficiently precise observations of the motions of the two stars could be made for a long enough period of years, it would become clear that the centre of mass was not at the focus of the apparent orbits and the inclination of the orbital plane to the line of light could be deduced from the fact that it must be at the focus of the true orbits; in practice this reduction is likely to be difficult. The problem is simplified if the eccentricities of the ellipses are small because, however a circular orbit is tilted, the maximum apparent dimension of the orbit is its diameter. In addition, if orbits appear almost circular, they must be approximately at right-angles to the line of sight.

Suppose that a double star system has a known parallax so that its distance from us is known and that in addition the two stars are observed to move in circular orbits (fig. 14). As the distance of the system is known, the apparent size of the orbit can be converted into the real size so that the radii r_1, r_2 of the two orbits are known. The centre of the orbits is the centre of mass of the system so that $M_1 r_1 = M_2 r_2$ or

$$M_1/M_2 = r_2/r_1, \tag{2.11}$$

where M_1 and M_2 are the masses of the two stars. Thus, as r_2 and r_1 are known, so is the ratio of the masses. Another relation between the masses can be obtained from Kepler's laws. These give a relation between the distance between M_1 and M_2, the sum of the masses and period of revolution P (the time taken for either star to describe its orbit). This relation is:

$$P^2 = 4\pi^2(r_1 + r_2)^3/G(M_1 + M_2). \tag{2.12}$$

Equations (2.11) and 2.12) enable the masses of both stars to be determined.

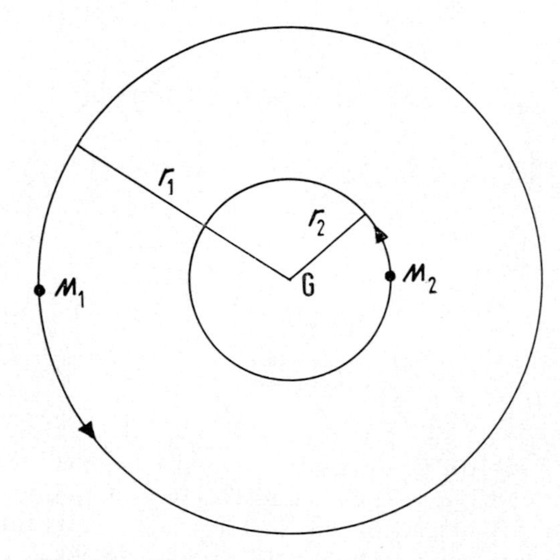

Fig. 14. Binary system with circular orbits.

Clearly, if the parallax of the binary system cannot be measured, only the apparent dimensions of the orbit will be known and in that case the masses of the individual stars cannot be found. However, the ratio of the masses will be known from equation (2.8). If M_2 is very much more massive than M_1, it will probably be impossible to determine r_2 accurately, even if the parallax can be measured. In that case equation (2.9) becomes approximately:

$$P^2 = 4\pi^2 r_1^{\,3}/GM_2 \tag{2.13}$$

and the mass of the principal component can be obtained.

Eclipsing binaries

It is also possible to obtain the masses of some close binary systems, but here the technique used is quite different. It involves the study of the spectra of the two components of the system. Before we discuss mass determination from the study of *spectroscopic binaries* we will first discuss *eclipsing binaries*. Such a binary is most simply described if it can be assumed that the less massive component moves in a circular orbit about the more massive component. If the plane of the orbit of a close binary system is almost such that it contains the line of sight from the Earth, it is possible that one star is eclipsed by the other at some stage during the orbit (fig. 15). The effect of this eclipse is to produce an

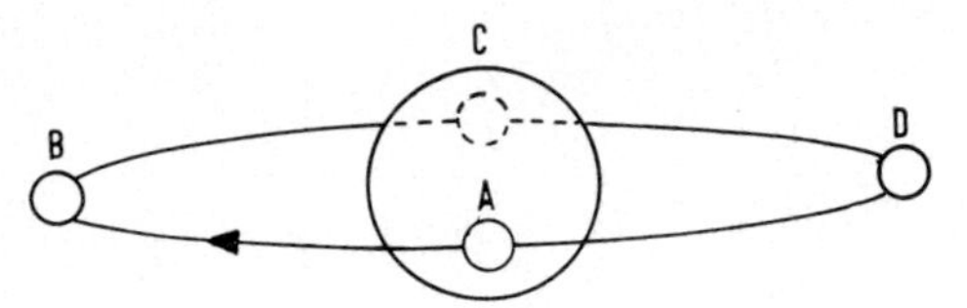

Fig. 15. Eclipsing binary. In position A the smaller star eclipses part of the larger star; at position C it is eclipsed.

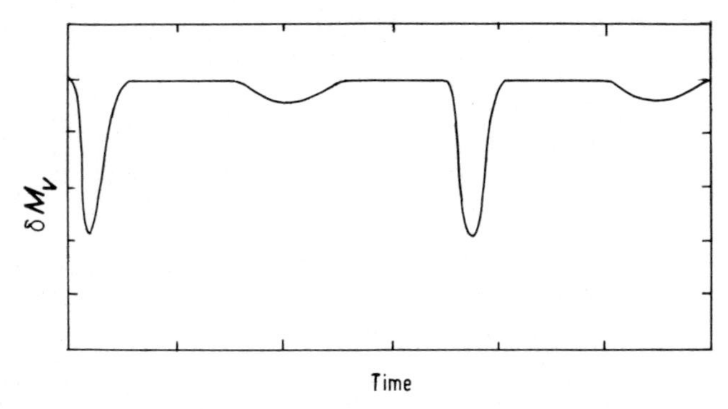

Fig. 16. The light curve of an eclipsing binary. Such a light curve would be obtained for the system of fig. 15 if the smaller star were very much hotter and the deep minimum would correspond to the eclipse of the small star.

apparent variation of the light output of the binary system as is shown in fig. 16. Quite a large amount of information can be obtained from such a light curve. In the first place the period of rotation of the secondary star around the primary can be found. The duration of the eclipses compared to the time between successive eclipses gives information about the radii of the stars compared to the size of the orbit. Finally, from the depth of the eclipses it may be possible to learn something about the angle of inclination of the plane of the orbit to the line of sight; the angle of inclination, i, is defined as the angle between the line of sight and the perpendicular to the orbital plane (fig. 17) and for eclipsing binaries this must be quite close to 90°.

Spectroscopic binaries

If the light from a close binary system is studied it may be possible to separate the spectral lines of the two components of the system. When the stars are at a phase in their orbits when one star has a component of velocity towards the earth and the other has a component of velocity away from the Earth, the spectral lines of the two stars are separated by the Doppler effect and it is possible to obtain the velocities of the two stars. If the angle of inclination were 90°, we would be able to observe the actual velocities of the two stars, but for an arbitrary angle of inclination only $v_1 \sin i$ and $v_2 \sin i$ can be observed, where v_1 and v_2 are the velocities of the two stars.

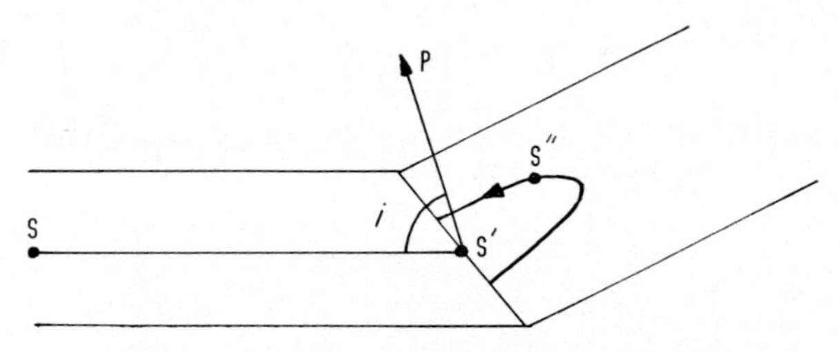

Fig. 17. The angle of inclination of a binary system. S represents the Sun and S′ and S″ the stars of a binary system, S′ being assumed (for simplicity) much more massive than S″. S′P is perpendicular to the orbital plane and the angle i, SS′P, is called the angle of inclination.

As each star takes the same time to describe the orbit, the velocity of either star is proportional to the radius of its orbit. In addition, from equation (2.11), the velocity is inversely proportional to the mass. Thus:

$$\frac{v_1 \sin i}{v_2 \sin i} \equiv \frac{v_1}{v_2} = \frac{r_1}{r_2} = \frac{M_2}{M_1}. \tag{2.14}$$

Thus the observation of $v_1 \sin i$ and $v_2 \sin i$ immediately gives a value for the mass ratio of the binary system. If v_1 and v_2 were known, we could obtain the radii of the orbits (assumed circular) from the observed period through the relations:

$$v_1 P = 2\pi r_1, \quad v_2 P = 2\pi r_2. \tag{2.15}$$

However, as only $v_1 \sin i$, $v_2 \sin i$ can be observed, only values of $r_1 \sin i$ and $r_2 \sin i$ can be obtained. Equation (2.12) can then be rewritten:

$$(M_1 + M_2) \sin^3 i = 4\pi^2 (r_1 + r_2)^3 \sin^3 i / GP^2, \tag{2.16}$$

where all the quantities on the right-hand side of equation (2.16) can be obtained from the observations. As spectroscopic binaries can be observed with arbitrary values of the angle of inclination, only a lower limit of the sum of the masses can be obtained from equation (2.16):

$$M_1 + M_2 \geqslant 4\pi^2(r_1 + r_2)^3 \sin^3 i/GP^2. \tag{2.17}$$

If a spectroscopic binary is also an eclipsing binary further progress can be made. In the first place, as the angle of inclination must be very close to 90° for the system to have eclipses, equation (2·17) can be replaced by:

$$M_1 + M_2 \approx 4\pi^2(r_1 + r_2)^3 \sin^3 i/GP^2, \tag{2.18}$$

and the masses of the two stars can be found from equations (2.14) and (2.18). In addition the true dimensions of the orbit are known from equations (2.15). The comparison of the time taken for eclipses with the period P now enables values to be obtained for the radii of the stars and this is one of the few ways in which stellar radii can be estimated. We now consider the problem of the measurement of radii.

Stellar radii

There are three different ways of obtaining values for the radii of stars. These involve direct measurement of stellar angular diameters by interferometry, study of eclipses as described above and use of the relation (2·7) between luminosity, radius and effective temperature.

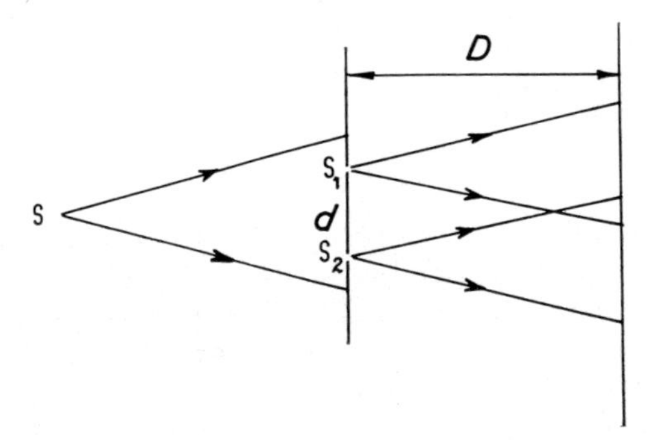

Fig. 18. Interferometry. Light from a source S falls on a screen containing two slits S_1 and S_2 and a pattern of bright and dark lines is formed on a second screen.

The simplest interferometric method can be described as follows. If light from a point source is allowed to fall on a screen containing two slits (fig. 18) and the light from the slits is then allowed to fall on another screen, a pattern of bright and dark lines is observed on this second screen. This is the phenomenon known as interference, which can easily be understood on the wave theory of light and which was one of the main experimental facts leading to the development of that theory. The distance between successive maxima and minima in the intensity is:

$$x = D\lambda/d, \tag{2.19}$$

where λ is the wavelength of the light, D is the distance between the two screens and d is the distance between the slits.

If the original source of the light is not a point source, or if the slits are too wide, the interference pattern may be destroyed. Thus, if the angular diameter of the source at the slit is θ (fig. 19), the interference

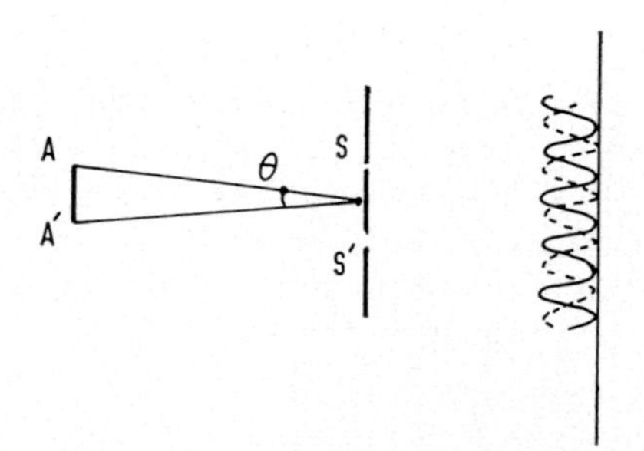

Fig. 19. If the source of fig. 18 is extended with diameter AA′, light from different parts of the source produces out of phase intensity patterns at the second screen as shown by the solid and dashed lines. If these patterns are sufficiently displaced, the overall interference pattern disappears.

fringes produced by light from different parts of the source overlap as shown in the figure and the interference fringes disappear when the distance between the slits is larger than

$$d = A\lambda/\theta, \tag{2.20}$$

where A is a number which is of the order of unity and which depends on the shape and density of illumination of the source. For a uniformly bright circular disk $A = 1{\cdot}22$ and A exceeds $1{\cdot}22$ for a disk which is darker at the edges, as the solar disk is observed to be. This increase of A for sources with their emission concentrated towards the centre is easy to understand. Such a source really behaves as a source of smaller angular diameter; from equation (2.20) reducing θ and increasing A give similar results.

Most stars have such small angular diameters that with practicable slit separations the interference pattern is observed and they appear as point sources. Some do have large enough angular diameters that it is possible to find the value of d at which the interference pattern disappears and that leads to an estimate of the angular diameter of the star provided a value can be obtained for A. Theoretical values of A are available for stars of different types. Once the angular diameter of a star is known, its linear diameter can be obtained provided that its distance is known.

Clearly this method will preferentially provide diameters of nearby stars of large diameter.

The estimation of radius using the properties of eclipsing binaries has been briefly described above. One point should be added. At the time the eclipse is occurring, the star is moving transverse to the line of sight, apart from any motion of the binary system as a whole towards or away from the Sun. At two other times in the star's motion it is moving either towards or away from the Sun and the difference in these velocities should enable the true orbital velocity of the star to be obtained. The time taken for the eclipse combined with this velocity then gives the radius of the star and this measurement does not require knowledge of the distance of the binary system.

The method of eclipses can be used in other ways. For example, a star may be in such a position in the sky that it apparently passes behind the Moon; this is known as a lunar occultation. In principle the time taken for the star's light to disappear from the moment of first contact gives a value of its angular diameter because the angular speed of the Moon is well known. This diameter can be converted into a true diameter if the star's distance is known. In actual fact the method of lunar occultation is not as simple as this because of diffraction of the star's light as it passes behind the Moon. However, the angular diameter of the star can be obtained from the properties of the diffraction pattern. Interferometric techniques and the method of lunar occultations have been much used by radio astronomers in their study of the angular diameters of cosmic radio sources.

Finally, radius can be estimated using equation (2.7):

$$L_s = \pi a c r_s^2 T_e^4,$$

but this is a much less reliable method than the other two. If stars really did radiate like black bodies the method would be straightforward. The distribution of radiation with wavelength would give the surface temperature of the star, T_*. The amount of radiation received on unit area of the Earth's surface in a given wavelength range could be measured and the radiation by unit area of a body at temperature T_* in the same wavelength range would be determined from the Planck function. The ratio of these two quantities would be r_*^2/d_*^2, where r_* is the radius of the star and d_* its distance. The radius could then be found if d_* were known.

In practice stars do not behave like black bodies. Attempts are made to measure the total radiation from the star falling on unit area of the Earth's surface by means of a bolometer. This is:

$$L_*/4\pi d_*^2 = a c r_*^2 T_{e*}^4/4d_*^2. \tag{2.21}$$

If the distance of the star is known, a value is then known for the product $r_*^2 T_{e*}^4$. If the star's radiation is not too different from that of a black body, it may be possible to estimate T_{e*} and hence r_*. However, as the uncertainty in r_* is the uncertainty in T_{e*}^2, this is only a very approximate method of estimating stellar radii.

It should finally be remarked that the Sun is the only star which appears as a disk rather than as a point of light and for the Sun a much more direct measurement of angular diameter and hence radius is possible.

Chemical composition; spectra

In the middle of the last century it was realized that the chemical elements possessed their own characteristic spectra. If heated to incandescence an element would emit radiation at various well defined frequencies; if the element was placed between the observer and a source of white light, it would absorb light of the same frequencies. Early in this century Bohr produced his model of the atom in which the electrons could exist in various orbits of definite energy around the positively charged nucleus and where energy was either emitted or absorbed when an electron moved from one energy level to another. This gave a natural explanation of the spectral lines of elements. Although later developments of the quantum theory have shown that Bohr's theory is not correct in detail, it is adequately true for our present purposes.

The characteristic spectral lines of many elements and sometimes molecules can be observed in the light received from stars. Sometimes they appear as *emission lines* where the light of a particular frequency is enhanced, but more often they appear as *absorption lines* where the emission from the star at the given frequency is less than that for neighbouring frequencies. Whether they are emission lines or absorption lines, their presence indicates that the element concerned is present in the outer layers of the star. As the distance radiation can travel inside a star before it is absorbed is very small compared to the radius of the star, direct information is only obtained about the chemical composition of the very outermost layers from which radiation escapes from the star. As we shall see later, we believe that the composition of the outer layers can sometimes be very unrepresentative of the star as a whole. Soon after the birth of the science of spectroscopy, it was found that most of the chemical elements were present in the outer layers of the Sun. In fact the existence of the element helium was first suggested by spectral lines from the Sun before it had been discovered on Earth.

Spectral types

When the spectra of a reasonable number of stars had been studied, it was found that the stars could conveniently be divided into a number of classes or *spectral types*. The division between the classes was not sharp but for most stars it was reasonably unambiguous. The spectral classes were based on which elements were most prominent in the spectra of the stars and these prominent elements varied considerably from star to star. In the Harvard classification the spectral types were denoted by capital letters A, B, C. . . . It was subsequently realized that some of the groups were superfluous and that a more meaningful order for those classes that remained was OBAFGKMRNS†. The main characteristics of these spectral types are shown in Table 2.

Originally it was thought that these observations were closely related to the chemical compositions of the stars and that the most prominent elements in the spectra were the most abundant elements in the stars. Later it was realized that the surface temperatures of the stars also played a vitally important role and the order OBA . . . is essentially an order of decreasing surface temperature.

O:	Ionized helium and metals, weak hydrogen.
B:	Neutral helium, ionized metals, hydrogen stronger.
A:	Balmer lines of hydrogen dominate, singly ionized metals.
F:	Hydrogen weaker, neutral and singly ionized metals.
G:	Singly ionized calcium most prominent, hydrogen weaker, neutral metals.
K:	Neutral metals, molecular bands appearing.
M:	Titanium oxide dominant, neutral metals.
R,N:	CN, CH, neutral metals.
S:	Zirconium oxide, neutral metals.

Table 2. Main features in the spectrum of stars of different spectral types.

The reason why temperature is very important in determining stellar spectra is as follows. If a particular spectral line is to be absorbed or emitted in a stellar atmosphere, there must be present atoms with electrons in the correct energy levels for the absorption or emission to occur. At low temperatures all of the atoms are in what are known as their *ground states*, with the electrons near to the nucleus. As the temperature rises, some electrons move into the higher *excited states* and later the atoms are ionized. Hydrogen has only one electron. The spectral lines of hydrogen which fall in the visible region of the spectrum are those which involve transitions down to and up from the first excited state (fig. 20). These lines are known as the Balmer series. These Balmer lines are only strong in stars with intermediate surface temperatures; at low temperatures all of the hydrogen is in the ground state and the only possible absorption lines are in the ultra-violet, while at high temperatures it is mainly ionized. As soon as it was realized that the spectral sequence was primarily a temperature sequence, it became clear that the variations of chemical composition from star to star were mainly slight. The approximate relation of surface temperature to spectral type is shown in Table 3. In the coolest stars of spectral types M–S

† The usual way of remembering the present order is through the mnemonic ‘ Oh be a fine girl kiss me right now, sweetheart ’.

relatively small abundance differences do have very important effects in the observed spectra as they determine exactly which molecules are formed.

Element abundances

Although the recognition that temperature plays a key role in the appearance of the spectra immediately demonstrated that there was

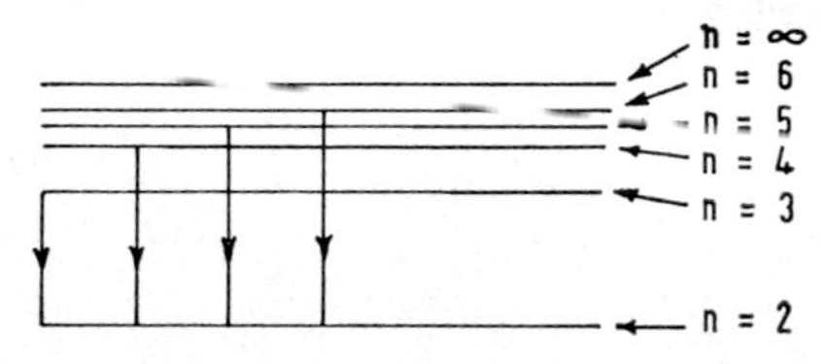

Fig. 20. The energy levels of the hydrogen atom. The vertical coordinate represents the energy difference between excited states and the ground ($n = 1$) state. The arrowed transitions to the first excited ($n = 2$) state represent emission of the Balmer series.

relatively little difference in the chemical composition of stars, there is still considerable difficulty in transforming the raw data of the observations into reliable chemical compositions. There are many factors involved, such as the detailed structure of the star's atmosphere and

Spectral type	O	B0	A0	F0	G0
T_e/K	50 000	25 000	11 000	7 600	6 000
Spectral type	K0	M0	M5	R,N,S	
T_e/K	5 100	3 600	3 000	3 000	

Each spectral type labelled by a capital letter is sub-divided into subclasses labelled by numbers, as in M5 above. The Sun has type G2.

Table 3. Effective temperature as a function of spectral type for main sequence stars

many atomic properties. As a result deduced chemical compositions are liable to be revised as the theory of the structure of stellar atmospheres is improved. Even in the case of the Sun, which can be observed much more thoroughly than any other star, the estimated abundances of the more abundant elements have recently been revised. However, the general character of the observations is reasonably clear. Hydrogen is the most abundant element, with only helium at all comparable. There is a gradual decline in abundances to higher atomic number with a decided local peak in the neighbourhood of iron and with several subsidiary peaks at higher atomic number. A schematic abundance curve is shown in fig. 21 which shows the logarithm of abundance as a function of atomic mass number.

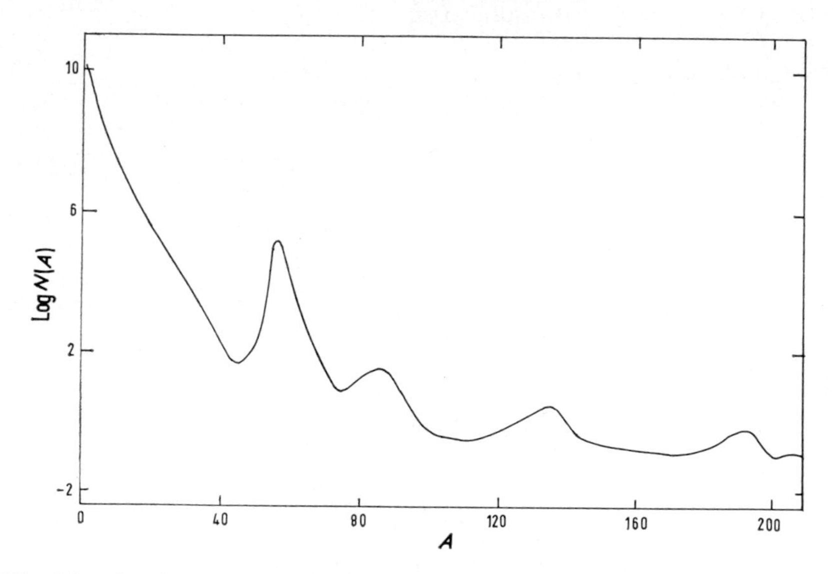

Fig. 21. A schematic abundance curve. A is the atomic mass number and $N(A)$ is the number of atoms with mass number A; the actual numbers are chosen so that there are 10^6 silicon atoms. The true abundance curve is much more irregular.

The origin of the elements

Although the abundance differences between stars are relatively slight, what differences there are, are very interesting and important. They lead to study of the problem of the origin of the chemical elements. It appears, for example, that the abundances of elements heavier than hydrogen and helium are lower in stars that were formed early in the life of the Galaxy than in stars which have been formed in the recent past. Indeed, it is possible that when the Galaxy was formed it contained no heavy elements and that they have been produced by nuclear reactions in stars in the lifetime of the Galaxy.

Later in the book we shall learn that we expect a succession of nuclear reactions to occur in stars, gradually building light elements into heavier elements. We shall also learn that the more massive a star is, the more rapidly it passes through its life history. Thus the earliest massive stars formed in the lifetime of the Galaxy could have produced the heavy elements which we find in the stars which have been formed more recently, provided that the heavy elements once formed were, at least in part, expelled into interstellar space so that they could be used in the formation of new stars. The problem of deciding whether a very simple initial chemical composition could have been changed into the composition shown in fig. 21, by nuclear reactions in stars in the lifetime of the Galaxy, is a very difficult one. It is outside the scope of the present book although most of the subject matter of this book is relevant to it.

General character of the observations

In the earlier part of this chapter we have discussed how the properties of stars can be observed, but we have not discussed the numerical values of mass, radius, luminosity etc. It is often convenient to express the properties of other stars in terms of those of the Sun. These are, with the suffix $\odot$ used to denote the Sun:

$$\left.\begin{aligned} M_\odot &= 1{\cdot}99 \times 10^{30}\ \mathrm{kg}, \\ L_\odot &= 3{\cdot}90 \times 10^{26}\ \mathrm{W}, \\ r_\odot &= 6{\cdot}96 \times 10^{8}\ \mathrm{m}, \\ T_{\mathrm{e}\odot} &= 5780\ \mathrm{K}. \end{aligned}\right\} \tag{2.22}$$

We can now state, in terms of the solar values, what ranges of values for M_s, L_s, r_s and T_e have been found in other stars. These are approximately

$$\left.\begin{aligned} 10^{-1}M_\odot &< M_s < 50M_\odot, \\ 10^{-4}L_\odot &< L_s < 10^{6}L_\odot, \\ 10^{-2}r_\odot &< r_s < 10^{3}r_\odot, \\ 2\times 10^{3}\ \mathrm{K} &< T_e < 10^{5}\ \mathrm{K}. \end{aligned}\right\} \tag{2.23}$$

The very high luminosities of exploding supernovae have been excluded from the above limits. It can be seen that there is quite a wide range in the values of all of these quantities, but that the luminosity range is definitely the most extreme. It should be stressed that these numbers refer to stars that have been observed and it is very likely indeed that stars exist with masses, radii and luminosities smaller than those shown in inequalities (2.23).

The Hertzsprung–Russell diagram

Although inequalities (2.23) give an idea of the range of stellar properties, more significant information is obtained by considering the correlation of these properties one with another. One such correlation is shown in a diagram known as the Hertzsprung–Russell diagram. Originally this diagram was a plot of absolute stellar magnitude in some wavelength range against spectral type. It was subsequently realized that variations in spectral type were equivalent to variations in surface temperatures and that logarithm of surface temperature could replace spectral type. However, as we have stated above, it is difficult to define surface temperature unambiguously and today the observer plots magnitude against colour index, say M_V against B–V, where M_V is an absolute magnitude corresponding to the apparent magnitude V. The resulting diagram is also known as a Colour–Magnitude diagram.

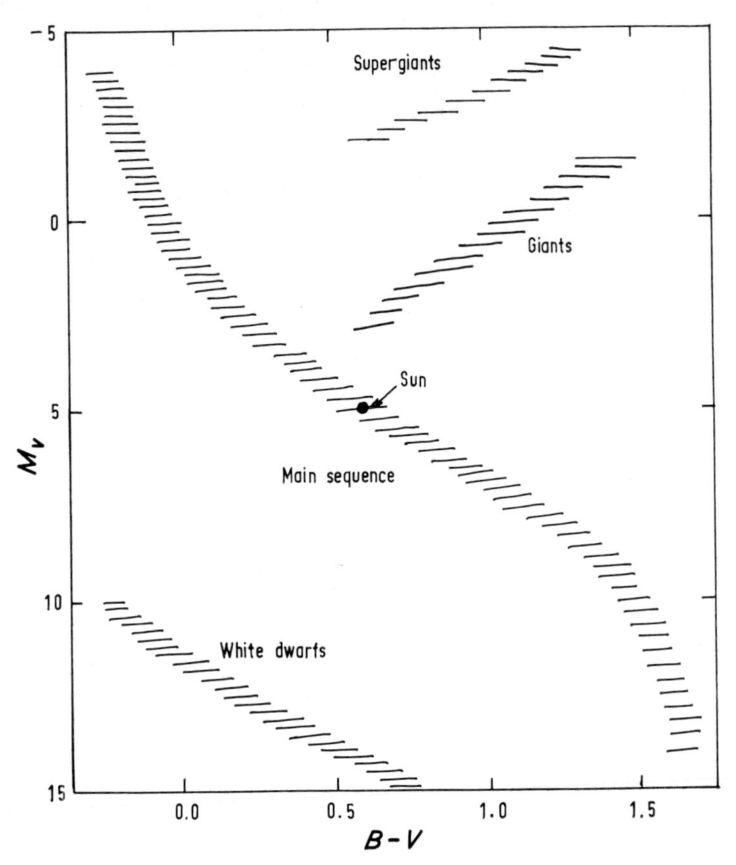

Fig. 22. The Hertzsprung–Russell diagram for nearby stars.

When the plot is made for the nearby stars of known distance, a diagram is obtained which is schematically as shown in fig. 22. The stars fall in four main regions. A band which contains the vast majority of the stars is called the main sequence and the other groups are called giants, supergiants and white dwarfs. The latter names arise because giants and dwarfs are found to have large and small radii respectively when these are known. It is a very important result that the stars do not lie uniformly over the whole of the HR diagram†. The fact that the stars are concentrated in particular regions with some correlation between magnitude and colour gives some hope that we shall be able to explain the observed properties of individual stars.

As stars must change with time, since they are radiating energy into space, we can ask whether evolution makes any significant difference to their positions in the HR diagram. Clearly at the very beginning and end of a star's life history we may expect its properties to be very different from its properties throughout most of its life, but do these properties change significantly during the major part of its life? In particular, about 90% of nearby stars are main sequence stars. Are *they* main sequence stars for the whole of their active life or are *all stars* main sequence stars for most of their life? It is with questions of this type that we shall be concerned in the later chapters of this book.

The mass–luminosity relation

If we now consider just those main sequence stars for which masses are known, it is found that there is a relationship between mass and luminosity which is illustrated in fig. 23. The more massive stars are the more luminous and the luminosity increases as a reasonably high power of the mass, $L_s \propto M_s^5$, in the steepest part of the curve. This is another relation which we must hope to understand theoretically. White dwarf stars, for which masses are known, do not obey the main sequence mass–luminosity relation. Although they are very much less luminous than main sequence stars, they have normal stellar masses. As we shall see in Chapter 6, there are good theoretical reasons for believing that red giants and supergiants, too, should not obey the main sequence mass–luminosity relation, but at present there is not even one really reliable giant mass known.

Cluster HR diagrams

As mentioned earlier, there are only a limited number of nearby stars for which direct measures of distance and hence of absolute luminosity can be made. If we had to rely on these observations, it is unlikely that we should be able to obtain a good theoretical understanding of stellar evolution. What is very useful is the existence of the star clusters which have been mentioned earlier; these include the large globular clusters with perhaps 100 000 or even 1 000 000 stars and the smaller galactic clusters. Both types of cluster are quite compact and physically bound together.

We do not have a direct estimate of cluster distances from the Earth. We can, however, obtain a very crude estimate of their distances by

† In what follows, we shall use the abbreviation HR diagram for Hertzsprung–Russell diagram.

assuming that the brightest cluster stars are generally similar to bright stars in the solar neighbourhood. As we shall soon see, we can improve on this first crude estimate of a cluster distance. The angular diameter of a cluster in the sky gives us a value for the ratio of the diameter of the cluster to its distance from us. For almost all clusters the angular diameter is small, which means that the cluster dimensions are very small compared to their distance from us. This means that all stars in any one cluster are essentially the same distance from us and in addition they probably suffer similar obscuration due to interstellar matter between us and the cluster.

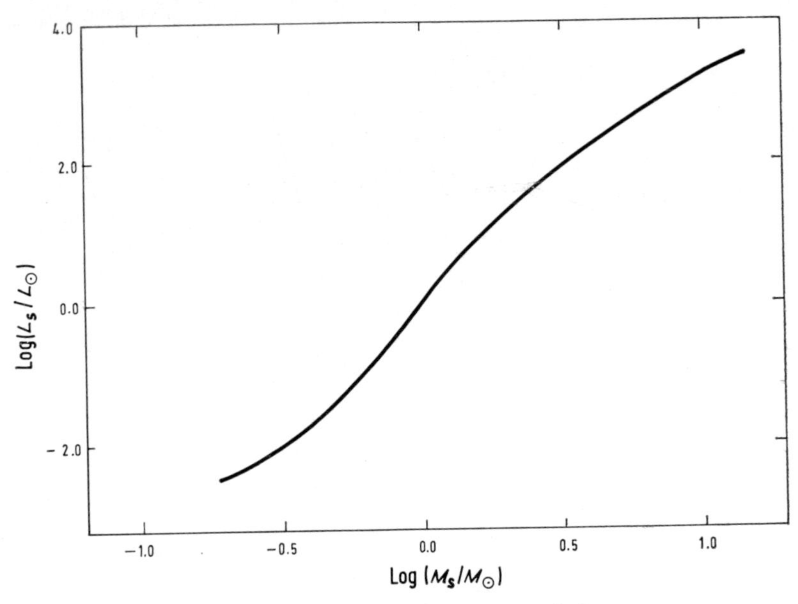

Fig. 23. The mass–luminosity relation.

Because the stars in a cluster appear to be physically associated, it is plausible that they were born close together at about the same time. If they were born out of the same cloud of interstellar gas, they may all have essentially the same chemical composition. Thus in trying to understand the properties of stars in a cluster we start by assuming that *all members of a cluster have the same age and chemical composition.* If this is so, the only reason why all of the stars in a cluster are not essentially identical is that they contain different quantities of matter. Thus *the principal factor differentiating stars in a cluster is mass.* This is not to say that other factors do not vary from star to star, but the hope is that these differences are relatively unimportant.

For a cluster of stars an HR diagram can be drawn in terms of apparent magnitude instead of absolute magnitude. When this is done it is found that cluster HR diagrams contain main sequences and giant branches, but that the spread in the diagrams is less than that in the HR diagram for nearby stars (fig. 22). In addition, in many cases there is a continuous transition between the main sequence and the giant branch with very few, if any, stars on the main sequence above the point where the giant branch joins it. The fact that the cluster HR diagrams are rather well defined adds to the hope that the cluster stars do form very homogeneous groups with only mass varying significantly from star to star in any cluster.

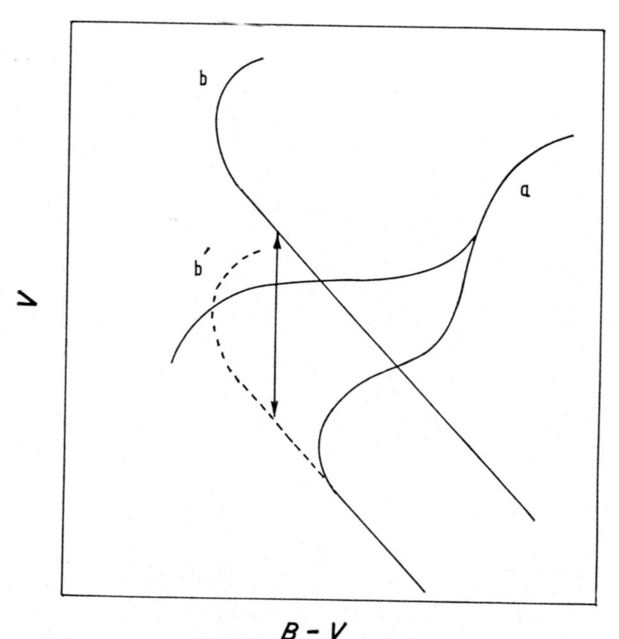

Fig. 24. Comparison of HR diagrams of two star clusters. Plotted in terms of apparent magnitude V, the main sequences of the two clusters fall in different positions. If it is assumed that they have the same absolute magnitude, the diagram of cluster b can be displaced vertically to b′, where its main sequence coincides with that of cluster a. This main sequence can then be assumed to have the same absolute magnitude as the main sequence in fig. 22.

It is obviously desirable that we should be able to convert the apparent luminosities of these cluster stars into absolute luminosities, as in any theoretical discussion it is the absolute luminosities which are predicted. Although distances to cluster stars cannot be measured directly, except perhaps for the nearest galactic clusters, there are fortunately ways in which the conversion from apparent to absolute magnitude can be made. Most of these methods rely on some interplay between theory and observation. The simplest approach involves the assumption that the stars of any given colour on a cluster main sequence have the same absolute magnitude as main sequence stars of the same colour in the solar neighbourhood. Because magnitude is related to the logarithm of luminosity, we can convert a cluster HR diagram to absolute magnitude by moving it vertically until its main sequence coincides with the main sequence of the nearby stars; the HR diagram is moved bodily and is not altered in shape in the conversion. This is illustrated in fig. 24. As the width of a cluster main sequence is usually less than the width of the

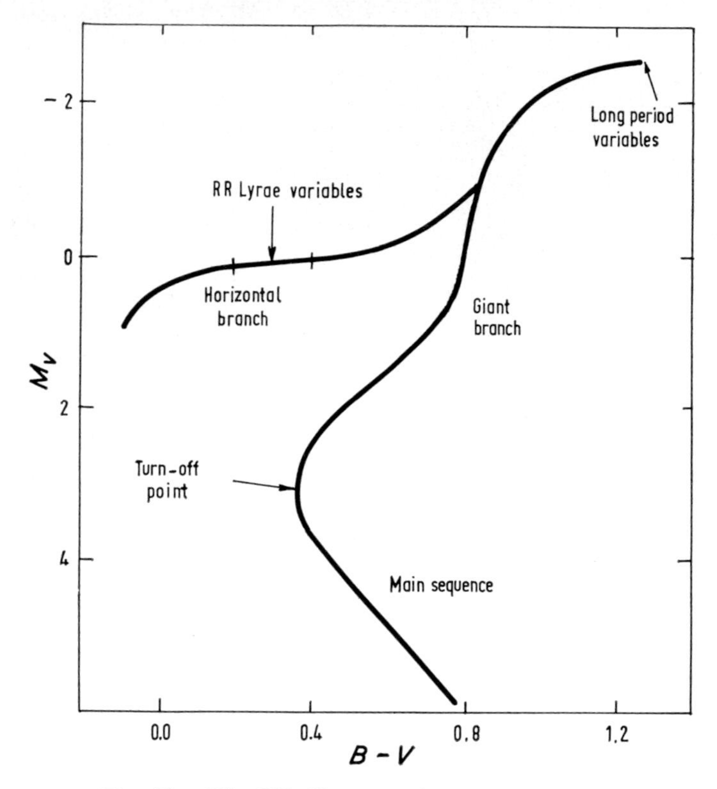

Fig. 25. The HR diagram of a globular cluster.

nearby stars main sequence, we could start by making an agreement between the cluster main sequence and the mean line of the sequence for nearby stars. In Chapter 5 we shall see how this procedure might be improved upon when we have some idea of what it is that causes a main sequence to have a finite width.

When the conversion from apparent to absolute magnitude has been made, the HR diagrams of globular and galactic clusters appear as in figs. 25 and 26. In fig. 25 is shown the diagram of one typical globular cluster and all globular cluster diagrams are reasonably similar to this one. There is a much greater variety in the HR diagrams of galactic clusters and four of these are sketched in fig. 26. An essential feature of all these diagrams, which has been mentioned above, is that there is a turn-off point from the main sequence. Below this point the cluster has a well-defined main sequence while above this point there are few stars on the main sequence.

Above we have been discussing the *stars in a cluster*, but it should be stressed that it is not really possible to label stars unambiguously as cluster stars. In the direction of a cluster, there are likely to be stars which lie between us and the cluster, and in the case of a nearby cluster

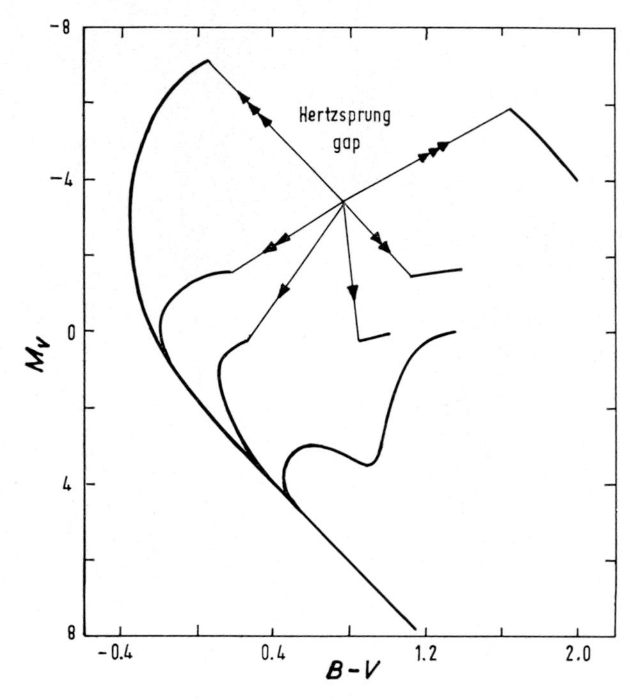

Fig. 26. The HR diagrams of several galactic clusters.

there may be stars visible which lie beyond the cluster. If we are only interested in doing statistics on the number of stars in a cluster without identifying individual cluster members, this can be done by counting stars in a region of the sky near to the cluster and by subtracting a similar number of stars per unit area from the count of cluster stars. Discovering exactly which stars are cluster members is more difficult. For some nearby clusters radial velocities and proper motions can be used to eliminate spurious members. In general more indirect methods must be used and there will always be some uncertainty.

The main features of the cluster HR diagrams shown in figs. 25 and 26 are as follows:

Globular clusters

1. They all have a main sequence turn-off in a similar position and a giant branch joining the main sequence at that point.
2. They have a horizontal branch running from near the top of the giant branch to the main sequence above the turn-off.
3. In many clusters there is a region of the horizontal branch which is populated only by stars of variable luminosity. These are known as RR Lyrae stars after the first star of their type to be studied and they will be discussed on page 42.

Galactic clusters

1. There is considerable variation in the position of the main sequence turn-off point, with the lowest being in about the same position as those of the globular clusters.
2. In many clusters there is a gap between the main sequence and the giant branch, known as the Hertzsprung gap.

Later in this book we shall see how the theory of stellar structure and evolution has given a general understanding of why globular and galactic cluster HR diagrams have the shape that they have.

Expanding stellar associations

As well as galactic and globular clusters, there are other groups of stars known as expanding associations. These are commonly groups of main sequence O and B stars of high luminosity, which are in the same region of the sky. When their properties are studied, it is found that they appear to be expanding from a common centre in such a way that their velocities are roughly proportional to their distance from the centre (fig. 27). In a typical case, if we extrapolate back their present velocities, they would all have been very close together a few million years ago. Expanding associations also occur containing stars which have irregular variations in their luminosity which are known as T Tauri stars and which will be discussed briefly on page 43 below. It is an obvious question to ask, if the stars in an expanding association were close together a few million years ago, where were they before that? It is believed that before that the stars did not exist as such and that in an expanding association we are observing the after-effects of a multiple star birth in the recent past history of the galaxy.

Special types of star

If a composite HR diagram is drawn up containing not only the nearby stars, but also the members of clusters whose distance has been obtained by indirect means, it looks schematically like fig. 28. In this diagram several special types of star are named and we now make a few remarks

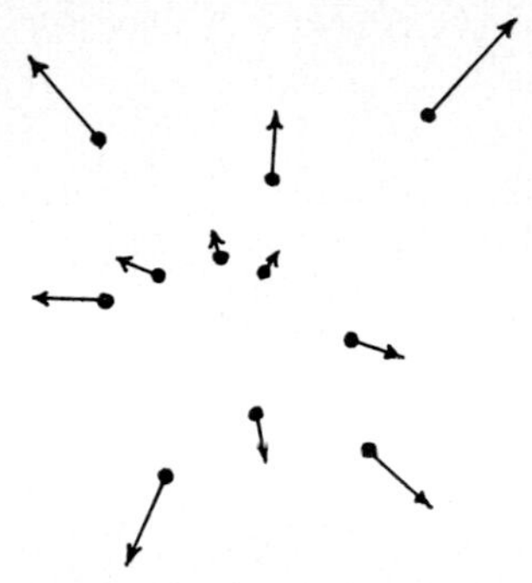

Fig. 27. The motion of stars in an expanding stellar association; the lengths of the arrows are proportional to the velocities of the stars.

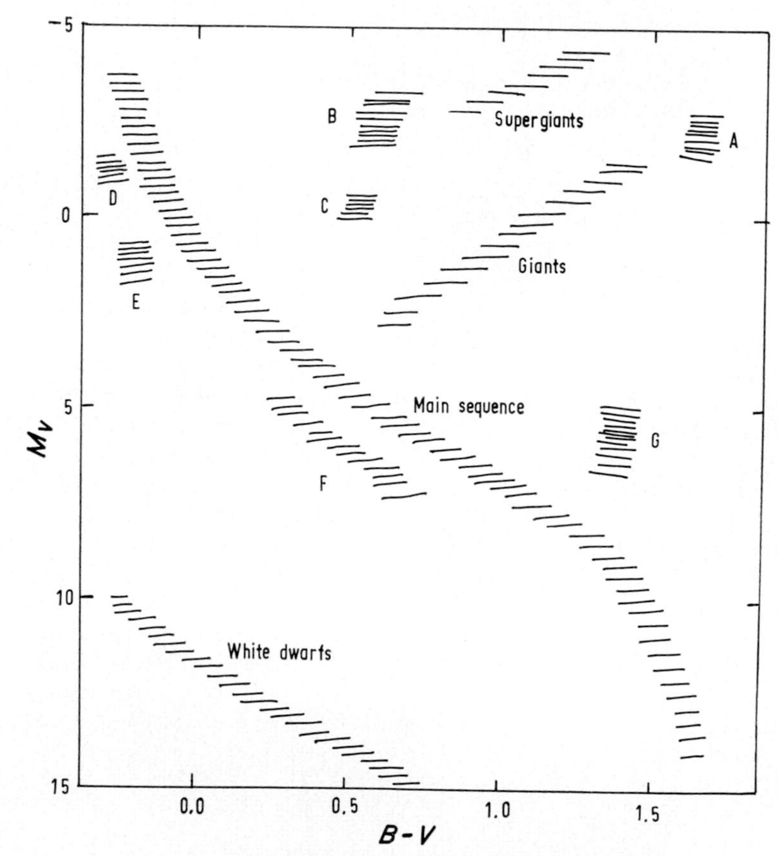

Fig. 28. A composite HR diagram including, A: long period (Mira) variables, B: cepheid variables, C: RR Lyrae variables, D: Wolf–Rayet stars, E: nuclei of planetary nebulae and old novae, F: sub-dwarfs, and G: T Tauri stars.

about each of these. It is first worth mentioning that groups of stars chosen because of characteristics other than luminosity and colour do form fairly compact groups in the HR diagram. We have already mentioned that some stars have variable luminosity and it is found that some regions of the HR diagram contain essentially no non-variable stars while other regions contain no variable stars.

White dwarfs

These are a particularly interesting group of stars about which we shall have more to say in Chapter 8. The best known white dwarf and the first to be discovered is the binary companion of the bright star Sirius. Although it is very much less luminous than Sirius, it is of approximately the same colour and it has a mass almost half as great and very similar to that of the Sun. Unless it has a very peculiar relationship between colour and effective temperature, this means that it must have a very small radius indeed. This follows from equation (2.7)

$$L_s = \pi acr_s^2 T_e^4,$$

with the luminosity known and the effective temperature estimated from the colour. All the evidence is that this is true and that *a match-box of white dwarf material would weigh a ton.* If this is true, material in white dwarfs must be at densities orders of magnitude higher than anything we meet or can hope to produce on Earth. This indicates that, if we wish to obtain a theoretical understanding of the structure of stars, we cannot rely on experiments in terrestrial laboratories to give us information about the behaviour of matter in all the conditions that occur in stellar interiors. In many cases we must rely on theory to predict the properties of matter in physical conditions which we cannot check experimentally.

Variable stars

There are various groups of stars whose light output varies in time and for which it can be shown that the variations are intrinsic to a single star and are not due to an eclipse in a binary system. Some of these stars are very regular and periodic with something near to a sinusoidal variation in light output, others have periodic, but irregular, light curves while others vary much more irregularly. Figure 29 shows the form of some typical light curves for periodic variables. For these the periods vary from a few hours to several hundred days. Several groups of regular variables are shown in fig. 28. These include the RR Lyrae stars with periods of a few hours, the cepheid variables typically with periods of about a week and the long period or Mira variables with periods of up to several hundred days†. The fact that each type of

† Each of these groups is named after the first star of the type discovered, RR Lyrae, δ Cephei and Mira Ceti.

variable is found in a rather compact region in the HR diagram seems very significant. This suggests that light variation is not an accident that can happen to any star, but that a specific combination of physical conditions must be required to make a star vary.

Cepheid variables

The cepheid variables have played a very important role in our understanding of the structure of the Galaxy and the Universe. It was soon discovered that they have a relation between their period and their mean luminosity (the period–luminosity relation) so that, if the period of a cepheid variable is known, so is its absolute magnitude. As the apparent magnitude can be observed an estimate of the distance of the star results. Cepheid variables have been used as *standard candles* in finding the distance to nearby galaxies in which cepheids can be observed.

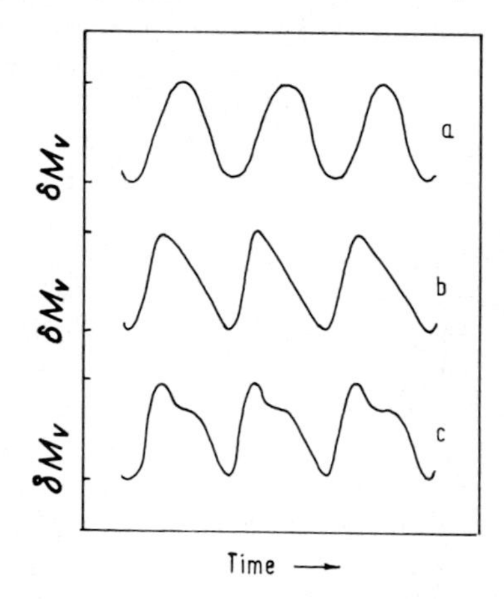

Fig. 29. Light curves for typical periodic variables. Some like a are very symmetrical, many like b have a more rapid rise to maximum light than decline to minimum and others like c have secondary humps on the light curve.

The period–luminosity relation is of course not an exact relation so that there is some uncertainty in the distances obtained. For a long time no cepheid variables were known in star clusters, but now a few are known in galactic clusters. They can be placed in the region of the Hertzsprung gap in the HR diagram and they give a second estimate of the distance to the cluster to supplement the fitting of the main sequence.

The RR Lyrae variables, which have shorter periods than the cepheids are found in large numbers in globular clusters. All RR Lyrae variables have a very similar luminosity. This means that the horizontal branches of all globular clusters should occur at about the same luminosity and making them coincide gives a second estimate of the distance of globular clusters. The RR Lyrae stars have lower luminosities than the cepheids, but very similar surface temperature and we shall see later that it is believed that the same physical process causes their variability. The Mira variables with very long periods also occur in globular clusters, but we shall not discuss them any further.

As well as regular variable stars there are some stars which vary irregularly in luminosity. Amongst these are the T Tauri variables which occur, as previously mentioned, in expanding associations and are also found just above the lower main sequence in some galactic clusters. They show quite large but irregular variations in luminosity and evidence of outflow of matter from their surfaces. We shall see in Chapter 5 that T Tauri stars are believed to be stars in the process of formation.

Novae and supernovae

Other stars which vary very violently and irregularly are the novae and supernovae. They suddenly increase in luminosity by many orders of magnitude. In both cases this increase in brightness is accompanied by an explosive loss of mass from the star. In the case of novae the loss of mass is relatively small and some stars have been novae more than once, but in the case of supernovae the explosion probably shatters the whole star. Only three supernovae have been observed in our Galaxy in the last thousand years, although many others have probably occurred and have been too far away to be discovered from the background of faint stars. Supernovae are so bright that they can be observed in quite distant galaxies. It is difficult to place the stars which become novae and supernovae in the HR diagram because their properties have probably not been studied before they explode. The remnants of supernovae are believed to be white dwarfs or neutron stars which will be discussed in Chapter 8. Post-novae can be placed in the HR diagram and they are marked in fig. 28.

Planetary nebulae

As well as novae and supernovae, there are also other stars which are observed to be losing mass at a non-catastrophic rate. In particular there are the planetary nebulae which are shown in fig. 28. These are stars which are surrounded by a sphere or spherical shell of gas which has almost certainly been ejected from the star at a previous stage, because the gas is observed to be expanding away from the star. They are called planetary nebulae because when looked at through a telescope they have a faint greenish disk which looks something like a planet. Stellar structure theory should eventually explain why some stars explode and why others lose mass less violently.

Sub-dwarfs

This name is used for all stars which lie significantly below the main sequence defined by the stars in the solar neighbourhood. Some of these stars have additional peculiarities, but the ordinary sub-dwarf apparently has a smaller abundance of elements other than hydrogen and helium than, say, the Sun. In Chapter 5 we shall see that this lower heavier element content may account for the position of the sub-dwarfs in the HR diagram.

Wolf–Rayet stars
These are very luminous blue stars which are apparently ejecting matter from their surfaces with velocities up to 10^6 m s^{-1}.

It can be seen that many of these special groups of stars are associated with variability of light output and/or stellar instability.

Stellar populations
In 1944, W. Baade introduced the concept that our Galaxy (and other galaxies) was composed of stars of two populations, *population I* and *population II*, and this concept has been important in all subsequent discussions of galactic structure and evolution and stellar evolution. In studying the nearest large galaxy to our own, the Andromeda galaxy M31, and its two companions, he showed that the composite HR diagram of the companions and of the central regions of M31 resembled that of a globular cluster (fig. 25). In particular the brightest stars in these systems were red supergiants. In contrast the HR diagram of the outer regions of M31 resembled that of a galactic cluster (fig. 26) with the brightest stars being blue main sequence stars.

He called the galactic cluster type stars population I stars and the globular cluster type stars population II stars. He found the central region and the halo region of our Galaxy was like the central regions of the Andromeda galaxy and was composed of population II stars while the disk was made up of population I stars. In addition he found that, from their position in the Galaxy, many of the special groups of stars discussed above could be classified as population I or II. Thus population I included cepheid variables, T Tauri stars, Wolf–Rayet stars and expanding associations, while the gas and dust of the Galaxy was also found in the region occupied by population I stars. Population II included RR Lyrae and Mira variables, planetary nebulae, sub-dwarfs and novae.

In Baade's original classification it was the position of a star in the Galaxy which primarily determined its population, but since then it has become clear that its place of origin is also important. Thus rapidly moving stars in the solar neighbourhood, which is mainly population I, may have been formed in the halo region of the Galaxy and they may be population II stars. Since Baade's original classification two things have become clear. Firstly, there is no really sharp distinction between two populations but there is a gradual transition between two extremes and secondly the main factors distinguishing the two populations are age and chemical composition. This will become clearer later in the book, but the general result is that population I stars are younger and have a larger abundance of heavy elements than population II stars. The fact that the gas and dust in the Galaxy are found in the regions populated by the population I stars means that it is possible for new stars to be formed in these regions.

Outline of contents of the following chapters

We shall see in the later chapters that the main task of stellar evolution theory to date has been to try to explain why the HR diagrams of globular and galactic clusters have the particular shape that they have and some of the results of this study will be mentioned here and will be

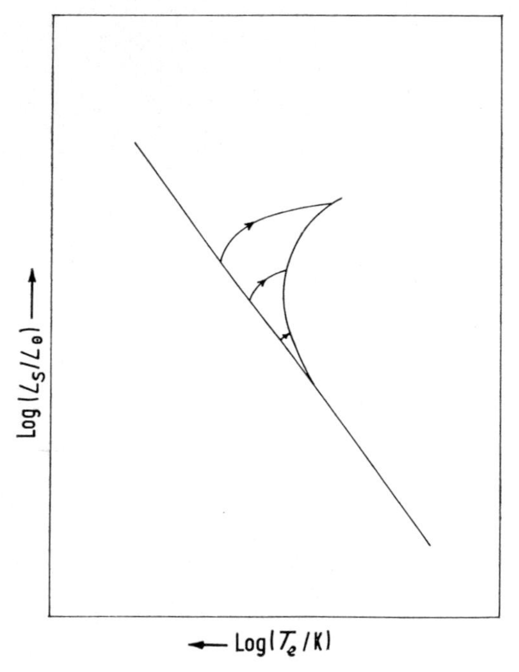

Fig. 30. Early stellar evolution. Individual stars evolve away from the main sequence along the arrowed tracks.

discussed in more detail later. It will be shown in Chapters 3 and 4 that the only physical processes capable of providing energy for stars to radiate for the length of time that they do radiate are nuclear fusion reactions which convert light elements into heavier elements. In Chapter 5 it will appear that the main sequence is populated by stars whose chemical composition is uniform, which are in the stage of burning hydrogen into helium in their interiors. This assumption leads to the prediction of a main sequence and of a main sequence mass–luminosity relation in good qualitative agreement with the observations.

In Chapter 6 it will be shown that, if the conversion of hydrogen into helium leads to the inside of the star becoming rich in helium while the outer regions still retain their initial chemical composition, the star's properties become those of a red giant. Furthermore, it is shown that, because the luminosity of a star increases very rapidly with its mass whilst its supply of nuclear fuel (for given chemical composition) only scales linearly with the mass, the more massive stars which are higher on the main sequence move into the giant region more rapidly than the less massive stars. This result is shown schematically in fig. 30, which shows how the evolution to the giant branch can lead naturally to the appearance of a turn-off point in the HR diagram. The older a system of stars, the lower the mass of star for which significant evolution has occurred and this suggests that, in observing galactic clusters with very different positions of turn-off point, we are observing systems of very different age.

All of this will be discussed more fully in the chapters which follow but, before we can do this, we must, in the next two chapters, discuss the basic principles governing the structure of a star and what are the important facts of physics which are needed for a discussion of the properties of stars.

Summary of Chapter 2

In this chapter we have discussed the main properties of stars which can, in principle, be deduced from observations. These are mass, radius, luminosity, surface temperature and chemical composition of the outer layers. Some estimate of surface temperature and chemical composition can be made for all stars that are near enough for a detailed study to be made of the distribution with wavelength of the light that they emit. The apparent brightness of a star can always be measured, but this can only be converted into a true luminosity for nearby stars whose distance from the Earth can be measured directly. Masses and radii can only be obtained for a very limited number of stars.

Progress in the study of stellar structure and evolution would have been very limited were it not for the existence of regularities in these properties. Thus for most stars there is a definite correlation between values of mass and luminosity and most stars lie in well-defined regions of the Hertzsprung–Russell diagram, which relates luminosity and surface temperature. Real progress in the theoretical interpretation of stellar properties is possible because many stars are members of clusters, which are more homogeneous groups of stars than an arbitrary set of nearby stars. Although it is not usually possible to observe all of the properties of individual cluster members, groups of stars, which are similar to nearby stars, can be found in clusters and this enables the distance to a cluster and the luminosity of its stars to be estimated. Throughout the remainder of this book we shall frequently use theoretical results to try to understand the HR diagrams of star clusters.

radius $r+\delta r$. We can now write down the condition that the net force on the element is zero; i.e. if the star is in equilibrium. Thus

$$P_{r+\delta r}\delta S - P_r\delta S + (GM_r\rho_r/r^2)\delta S\delta r = 0. \tag{3.2}$$

Provided an infinitesimal element is being considered we can write:

$$P_{r+\delta r} - P_r = (\mathrm{d}P_r/\mathrm{d}r)\delta r. \tag{3.3}$$

If equations (3.3) and (3.2) are combined, we obtain:

$$\frac{\mathrm{d}P_r}{\mathrm{d}r} = -\frac{GM_r\rho_r}{r^2}$$

In what follows we shall omit the suffixes r on P, M and ρ and will understand these symbols to mean the pressure at radius r, the mass contained within radius r and the density at radius r. Thus the above equation is written:

$$\frac{\mathrm{d}P}{\mathrm{d}r} = -\frac{GM\rho}{r^2}. \tag{3.4}$$

Equation (3.4) is known as the *equation of hydrostatic support.*

In equation (3.4) the three quantities M, ρ and r are not independent since the mass contained within a sphere of radius r is determined by the density of the material at points within radius r. A relation between M, ρ and r can be obtained as follows. Consider the mass of a spherical shell between radii r and $r+\delta r$ (fig. 32).

The mass of this shell is approximately $4\pi r^2\rho\delta r$, provided that δr is small. The mass of the shell is also the difference between $M_{r+\delta r}$ and M_r, which for a thin shell can be written:

$$M_{r+\delta r} - M_r = (\mathrm{d}M/\mathrm{d}r)\delta r.$$

Then, equating the two expressions for the mass of the spherical shell, we obtain:

$$\frac{\mathrm{d}M}{\mathrm{d}r} = 4\pi r^2\rho. \tag{3.5}$$

Equation (3.5) may be written alternatively as:

$$M_r = \int_0^r 4\pi r'^2\rho_{r'}\,\mathrm{d}r'. \tag{3.6}$$

We have obtained two of the equations of stellar structure. These are two differential equations for the three quantities P, M and ρ in terms of r. It is clear that we require a further relation between them if we can hope to determine them all. There is one fairly obvious type of relation, the equation of state of the stellar material, analogous to the equation of state of an ideal gas. This will relate the pressure and density but it will also in general introduce another quantity, the temperature T. We shall therefore still require at least one more equation. Before discussing these additional equations, it is possible to obtain some useful general information about the structure of stars on the basis of equations (3.4) and (3.5) alone. First, however, we will discuss when the two basic assumptions of this chapter are likely to be valid.

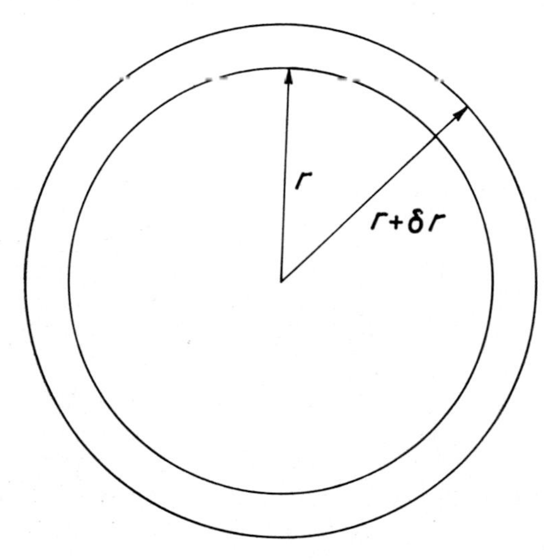

Fig. 32.

Accuracy of hydrostatic assumption

In the derivation of equation (3.4) it has been assumed that the forces acting on any element of material in a star are exactly in balance. As we shall see later, during its life history a star undergoes periods of radial expansion and contraction and at these times equation (3.4) is not accurately true. In these circumstances we can generalize equation (3.4) as follows. The net force acting on an element must be equated to the product of its mass and acceleration. If a is defined to be the acceleration in the *inward* radial direction a term $\rho_r a \delta S \delta r$ must be introduced on the right-hand side of equation (3·2) and equation (3.4) becomes:

$$\rho a = \frac{GM\rho}{r^2} + \frac{\partial P}{\partial r}, \tag{3.7}$$

where the partial derivative, $\partial P/\partial r$, is used as P is now a function of both r and t.

We can now estimate what would happen if the two terms on the right-hand side of equation (3.7) were slightly out of balance. Suppose that their sum is a fraction λ of the gravitational term so that the inward radial acceleration is a fraction λ of the acceleration due to gravity ($g \equiv GM/r^2$). If the element starts from rest with this acceleration, its inward displacement (s) will be given by:

$$s = \tfrac{1}{2}\lambda g t^2. \tag{3.8}$$

The radius will decrease by 10%, for example, in the time

$$t = \sqrt{\left(\frac{r}{5\lambda g}\right)}. \tag{3.9}$$

At the surface of the sun $r \simeq 7 \times 10^8$ m and $g \simeq 2{\cdot}5 \times 10^2$ m s^{-2} so that

$$t \simeq 10^3/\lambda^{1/2}\text{s}. \tag{3.10}$$

Since geological evidence concerning the ages of the radioactive elements in the Earth's crust and of fossils suggests that the properties of the Sun have not changed significantly for at least 10^9 years (3×10^{16} s), we can see that at present λ can be no greater than 10^{-27} so that equation (3.4) must be true to a very high degree of accuracy indeed. To put this another way, if the force of gravitation were not resisted by the pressure gradient of the solar material so that $\lambda = 1$, the radius of the Sun would change significantly in an hour. In the previous chapter we have mentioned that stars do exist in which significant changes occur in hours or days. These include novae, supernovae and some types of variable star. For such stars equation (3.4) must be replaced by equation (3.7).

From equation (3.8) we can obtain an expression for what we earlier called the dynamical time-scale of a star. If we put $s = r$ and $\lambda = 1$ we obtain an estimate of how long it would take the star to collapse completely if pressure forces were negligible. This we define to be the dynamical time, t_d, and it is given by:

$$t_d = (2r^3/GM)^{1/2}. \tag{3.11}$$

Validity of assumption of spherical symmetry

One reason why stars are not accurately spherically symmetrical is that they rotate. Rotating bodies, which are composed of liquids or gases, are flattened at the poles. There is a difference between the polar and equatorial diameter of the Earth which presumably dates from the time when the Earth was molten. In the case of most stars this effect is very small but there are some stars which rotate very rapidly; these are seriously non-spherical and we cannot discuss them in detail in this book.

We can now make an approximate estimate of the importance of rotation in determining the shape of the star. Consider an element of material of mass m near the surface of the star at the equator (fig. 33). In addition to the gravitational and pressure forces, the element will be acted on by an outward force $m\omega^2 R$, where ω is the angular velocity of the star and R is the equatorial radius. This force will be of negligible importance compared to the gravitational force and will therefore not cause a serious departure from spherical symmetry, provided that

$$m\omega^2 R/(GMm/R^2) \ll 1$$

or

$$\omega^2 \ll GM/R^3. \tag{3.12}$$

This expression is closely related to equation (3.11) and we can say that rotation will have only a slight influence on the structure of a star, provided that the rotation period is very large compared to the dynamical time, t_d.

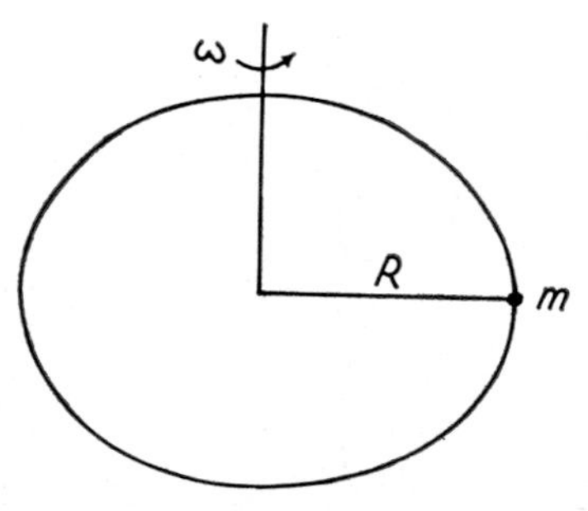

Fig. 33.

In the case of the Sun, the effects of rotation are very small. The Sun rotates once in about a month so that $\omega \simeq 2{\cdot}5 \times 10^{-6}\ \mathrm{s}^{-1}$ and $\omega^2 r^3/GM \simeq 2 \times 10^{-16}$, which suggests that we can reasonably neglect departures from spherical symmetry due to rotation. The rotation of the Sun is interesting and not fully understood. It does not rotate as a solid body and the equatorial regions rotate more rapidly than the polar regions. In contradiction to the estimate that we have made above, one observer claims to have measured a slight flattening of the Sun. He claims that this implies that the Sun's interior rotates much more rapidly than its exterior, which would modify the rough estimate we have made above. At present there is no general agreement about this result and its interpretation.

We now consider stars in which departures from equilibrium and spherical symmetry are unimportant and discuss some consequences of equations (3.4) and (3.5).

Minimum value for central pressure of a star

Use of equations (3.4) and (3.5) alone, with no knowledge of the type of material of which a star is composed, enables us to find a minimum value for the central pressure of a star whose mass and radius are known. If we divide equation (3.4) by equation (3.5) we obtain:

$$\frac{\mathrm{d}P}{\mathrm{d}r}\bigg/\frac{\mathrm{d}M}{\mathrm{d}r} \equiv \frac{\mathrm{d}P}{\mathrm{d}M} = -\frac{GM}{4\pi r^4}. \tag{3.13}$$

Equation (3.13) can now be integrated with respect to M between the centre of the star and its surface to give:

$$-\int_0^{M_s} \frac{dP}{dM} dM = P_c - P_s = \int_0^{M_s} \frac{GM}{4\pi r^4} dM, \tag{3.14}$$

where here, and in what follows, the suffixes c and s refer to the centre and surface of the star. Thus M_s is the total mass of the star, P_c is the central pressure and P_s the surface pressure.

We can now obtain an underestimate of the integral on the right-hand side of equation (3.14). At all points inside the star r is less than r_s and hence $1/r^4$ is greater than $1/r_s^4$. This means that

$$\int_0^{M_s} \frac{GM}{4\pi r^4} dM > \int_0^{M_s} \frac{GM dM}{4\pi r_s^4} = \frac{GM_s^2}{8\pi r_s^4}. \tag{3.15}$$

Equations (3.14) and (3.15) can now be combined to give:

$$P_c > P_s + GM_s^2/8\pi r_s^4 > GM_s^2/8\pi r_s^4. \tag{3.16}$$

For the Sun we know accurate values of M_s and r_s and these can be inserted into inequality (3.16) to give:

$$P_{c\odot} > 4{\cdot}5 \times 10^{13} \text{ N m}^{-2} \tag{3.17}$$

or

$$P_{c\odot} > 4{\cdot}5 \times 10^{8} \text{ atmospheres.} \tag{3.18}$$

This is a remarkably powerful result which requires no knowledge of the chemical composition or physical state of the solar material. Clearly, however, it gives some information about the possible physical state of the material at the centre of the Sun. It may seem surprising, in view of the very high value of this pressure, that we believe that the solar material is gaseous. As we shall see shortly, it is not an ordinary gas.

For stars other than the Sun inequality (3.16) can be rewritten as follows:

$$P_c > (GM_\odot^2/8\pi r_\odot^4)(M_s/M_\odot)^2(r_\odot/r_s)^4.$$

Then, using the solar values in the first expression in brackets:

$$P_c > 4{\cdot}5 \times 10^{13}(M_s/M_\odot)^2(r_\odot/r_s)^4 \text{ N m}^{-2}. \tag{3.19}$$

The virial theorem

A further consequence of the fundamental equations (3.4) and (3.5) can be found by integrating the equations over the entire volume of the star. From equations (3.4) and (3.5) we can obtain:

$$4\pi r^3 dP = -4\pi r GM\rho dr = -(GM/r)dM. \tag{3.20}$$

Integrating (3.20) over the whole star:

$$3\int_{P_c}^{P_s} V dP = -\int_0^{M_s} (GM/r) dM, \tag{3.21}$$

where V is the volume contained within radius r. Integrating the left-hand side of (3.17) by parts, the equation can be written:

$$3\Big[PV\Big]_{c}^{s} - 3\int_{0}^{V_s} P\mathrm{d}V = -\int_{0}^{M_s} (GM/r)\mathrm{d}M. \qquad (3.22)$$

The integrated part vanishes at the lower limit of integration because $V_c = 0$. The term on the right-hand side of equation (3.22) is the negative gravitational potential energy of the star (i.e. apart from the minus sign it is the energy released in forming the star from its component parts dispersed to infinity) and we denote this by the symbol Ω. Noting that $\mathrm{d}M = \rho \mathrm{d}V$, equation (3.22) can be written:

$$4\pi r_s^{\,3} P_s = 3\int (P/\rho)\mathrm{d}M + \Omega. \qquad (3.23)$$

If the star were surrounded by a vacuum, its surface pressure would be zero and the left-hand side of equation (3.23) could be put equal to zero. In fact the surface pressure of a star will not be zero but it will be many orders of magnitude smaller than the central pressure or the mean pressure in the interior. This means that the term on the left-hand side of equation (3.23) is very small compared to either of the terms on the right-hand side and it can usually be neglected and (3.23) can be written in the approximate form:

$$3\int (P/\rho)\mathrm{d}M + \Omega = 0. \qquad (3.24)$$

Equation (3.24) is usually known as the *Virial Theorem* and we shall use it frequently later in the book.

The physical state of stellar matter

In the early years of the study of stellar structure there was much discussion about the physical state of matter in stars. It was thought that the stars could not be solid because their temperatures were so high and that they could not be gaseous because their mean densities were too high. It is now believed that they are composed of an almost perfect gas in most circumstances. The perfect gas is, however, unusual in two respects.

The most important respect is that the stellar material is an ionized gas or *plasma*. The temperature inside stars is so high that all but the most tightly bound electrons are separated from the atoms. This makes possible a very much greater compression of the stellar material without deviation from the perfect gas law because a nuclear dimension is 10^{-15} m compared with a typical atomic dimension of 10^{-10} m. The word plasma is the name given to a quantity of ionized gas. It has been recognized in recent years that a plasma can be regarded as a fourth state of matter and that most of the material in the universe is in this fourth state. It differs from an ordinary gas because the forces between electrons and ions have a much longer range than the forces between neutral atoms.

The second important difference between most laboratory conditions and conditions in stars is that radiation is in thermal equilibrium with matter in stellar interiors, and its intensity is governed by Planck's law (2.5). Just as the particles in a gas exert a pressure which can be calculated from the kinetic theory of gases by considering collisions of particles with an imaginary surface in the gas, the photons in a Planck distribution exert a pressure known as radiation pressure. At one time it was thought that radiation pressure was of comparable importance to gas pressure in ordinary stars. It is now realized that although there are some exceptional stars in which radiation pressure is of vital importance, it is only of marginal significance in most stars.

From the kinetic theory of gases, the pressure of a perfect gas can be shown to have the form:

$$P_{\mathrm{gas}} = nkT, \tag{3.25}$$

where n is the number of particles per cubic metre and k is Boltzmann's constant ($1{\cdot}38 \times 10^{-23}$ J K^{-1}). This expression for the pressure can be made to correspond with the usual form for Boyle's law as follows. If we consider a mass of gas $\mathscr{M}$, of molecular weight m, which occupies a volume v, its pressure is given by:

$$P_{\mathrm{gas}}v = \frac{\mathscr{M}}{m}RT = \frac{\mathscr{M}}{m}N_{\mathrm{A}}kT, \tag{3.26}$$

where R is the gas constant ($8{\cdot}31$ J mol^{-1} K^{-1}), N_{A} is Avogadro's number ($6{\cdot}02 \times 10^{23}$ mol^{-1}) and $k = R/N_{\mathrm{A}}$. If we consider a cubic metre of gas and note that $n(\equiv \mathscr{M}N_{\mathrm{A}}/m)$ is the number of particles in a cubic metre, (3.25) results. The corresponding expression for radiation pressure is:

$$P_{\mathrm{rad}} = \frac{1}{3}aT^4, \tag{3.27}$$

where a is the radiation density constant ($7{\cdot}55 \times 10^{-16}$ J m^{-3} K^{-4}).

Minimum value of mean temperature of a star

In the Sun and many other stars radiation pressure is negligible. We shall now attempt to justify this statement as follows. We shall assume that stars are composed of a perfect gas with negligible radiation pressure. We shall then use the Virial Theorem to obtain a *lower bound* to the mean stellar temperature. We already know something about stellar densities from observations of masses and radii. At the temperatures and densities found, it appears that the stellar material would indeed be gaseous and radiation pressure would be negligible.

Consider the two terms in the Virial Theorem:

$$3\int (P/\rho)\,\mathrm{d}M + \Omega = 0.$$

The magnitude of the gravitational potential energy has a lower bound in terms of the total mass and radius of the star. Thus:

$$-\Omega = \int_0^{M_s} \frac{GM\,\mathrm{d}M}{r}.$$

In this integral r is less than r_s everywhere within the star so that $1/r$ is greater than $1/r_s$. Thus:

$$-\Omega > \int_0^{M_s} \frac{GM\,\mathrm{d}M}{r_s} = \frac{GM_s^2}{2r_s}. \tag{3.28}$$

If the star is assumed to be a perfect gas with negligible radiation pressure, the other term in the Virial Theorem can be written:

$$3\int_0^{M_s} \frac{P}{\rho}\,\mathrm{d}M = 3\int_0^{M_s} \frac{kT}{m}\,\mathrm{d}M = \frac{3k}{m}M_s\bar{T}, \tag{3.29}$$

where $\rho = nm$ so that m is now the average mass of the particles in the stellar material and $\bar{T}$ is a mean temperature defined by:

$$M_s\bar{T} = \int_0^{M_s} T\,\mathrm{d}M. \tag{3.30}$$

Combining equations (3.24) and (3.30) and inequality (3.28):

$$\bar{T} > GM_s m/6kr_s. \tag{3.31}$$

For the Sun we can insert values of the mass and radius into inequality (3.31) and can express the mean particle mass in terms of the mass of the hydrogen atom ($m_H = 1.67 \times 10^{-27}$ kg) to obtain:

$$\bar{T}_\odot > 4 \times 10^6 (m/m_H)\ \mathrm{K}. \tag{3.32}$$

To obtain a numerical value for this lower limit to the mean temperature, we need a value for m/m_H. As we have learnt in Chapter 2, hydrogen is the most abundant element in stars and for fully ionized hydrogen $m/m_H = \frac{1}{2}$, as there are two particles, one proton and one electron, for each hydrogen atom. For any other element whether fully ionized or not the value of m/m_H is greater; this will be discussed in detail in Chapter 4. Thus we can certainly write:

$$\bar{T}_\odot > 2 \times 10^6\ \mathrm{K}. \tag{3.33}$$

This is a very high temperature by terrestrial standards and it is also very much higher than the observed surface temperatures of the Sun and other stars. We also have an estimate of the mean density of the Sun through the relation:

$$\bar{\rho}_\odot = 3M_\odot/4\pi r_\odot^3 \simeq 1{\cdot}4 \times 10^3\ \mathrm{kg\ m^{-3}}. \tag{3.34}$$

It is now possible to verify that material with the mean density of the Sun at the mean temperature of the Sun will be a highly ionized gas. The mean density of the Sun is only a little higher than that of water and other ordinary liquids and such liquids turn to gases at temperatures much lower than that given by inequality (3.33). In addition at such a temperature, the average kinetic energy of the particles is higher than the energy required to remove many bound electrons from atoms and the gas will thus be highly ionized.

We can also estimate the importance of radiation pressure at a typical point in the Sun. Thus from equations (3.25) and (3.27):

$$\frac{P_{\text{rad}}}{P_{\text{gas}}} = \frac{aT^3}{3nk}. \tag{3.35}$$

With $T \simeq \bar{T} \simeq 2 \times 10^6$ K and $n \simeq 2\bar{\rho}/m_{\text{H}} \simeq 2 \times 10^{30}$;

$$\frac{P_{\text{rad}}}{P_{\text{gas}}} \simeq 10^{-4}. \tag{3.36}$$

In this calculation we have underestimated $\bar{T}$ but it certainly appears that radiation pressure is unimportant at an average point in the Sun. It should be stressed that these last two discussions are specific to the Sun. Although many other stars are composed of a near perfect gas with negligible radiation pressure, there are stars in which the gas is highly imperfect and others in which radiation pressure is important.

The source of stellar energy

So far we have really only considered the dynamical properties of a star. However, perhaps the most important property of a star is that it continuously radiates energy into space and we must concern ourselves with the origin of that energy and how it is transported from its place of origin to the surface of the star. Let us first consider the origin of this energy, where, of course, we do not mean the appearance of energy from nothing, but its conversion from another form in which it is not immediately available for the star to radiate. Once again we take the Sun as an example. The Sun radiates energy at a rate of 4×10^{26} J s^{-1} (4×10^{26} W). Using Einstein's relation between mass and energy, $E = mc^2$, this means that the Sun is losing mass at the rate of 4×10^9 kg s^{-1}. By studying the radioactive elements in the Earth's rocks and their decay products, it is possible to estimate how long the rocks have been solid. Study of the fossils in the rocks indicates how long living things have been present on Earth. These studies show that the Sun's luminosity cannot have changed significantly in the last few thousand-million years and in that time the total mass loss must have been about $2 \times 10^{-4} M_\odot$.

What can have been the source of this energy? Perhaps the simplest idea is that the Sun became very hot at some time in the remote past, perhaps it was created very hot, and has since been cooling down. We can test the plausibility of this by asking for how long the present thermal energy content of the Sun could supply its present rate of energy loss. Another possibility which was seriously considered when the structure of stars was first studied was that the Sun was slowly contracting with a consequent release of gravitational potential energy and that this energy was converted into the radiation which escaped from the surface.

The thermal energy and the gravitational energy of a star composed of a perfect gas are very closely related. In a perfect gas the total thermal energy is obtained by multiplying the number of particles by the number of degrees of freedom, n_f, possessed by each particle and by $kT/2$. Thus the thermal energy per unit volume is $nn_f kT/2$. The number of degrees of freedom n_f is related to the ratio of specific heats γ of the material by $\gamma = (n_f + 2)/n_f$, where γ is the ratio of the specific heat at constant pressure to the specific heat at constant volume. Using the expression (3·25) for the pressure of a perfect gas and introducing the thermal energy per kilogramme, u, instead of the thermal energy per unit volume,

$$u = P/(\gamma - 1)\rho. \tag{3.37}$$

The Virial Theorem (3.24) can then be written

$$3(\gamma - 1)U + \Omega = 0 \tag{3.38}$$

for a perfect gas with negligible radiation pressure, where U is the total thermal energy of the star. As mentioned earlier, the material inside a star is highly ionized. A fully ionized gas is a monatomic gas for which the value of γ is 5/3; for such a value of γ equation (3.38) can be written:

$$2U + \Omega = 0. \tag{3.39}$$

Thus for such a star the negative gravitational energy is just equal to twice the thermal energy.

It is clear from equation (3.39) that the time for which the present thermal energy of the Sun can supply its radiation and the time for which the past release of gravitational potential energy could have supplied its present rate of radiation differ only by a factor of two and to get an approximate idea of the time only one need be considered. The total release of gravitational potential energy is of order $(GM_\odot^2/r_\odot)$ J and this would have been sufficient to provide radiant energy at a rate $L_\odot$ W for a time:

$$GM_\odot^2/r_\odot L_\odot \simeq 10^{15}\ \mathrm{s} \simeq 3 \times 10^7\ \text{years}. \tag{3.40}$$

This means that if the Sun's radiation were supplied by either contraction or cooling (we can consider either since the factor of 2 is insignificant), then it would have changed substantially in the last ten million years, while geologists tell us that it can hardly have altered in a time a hundred times longer. The time

$$t_{th} \equiv GM_s^2/r_s L_s, \tag{3.41}$$

for any star, is the thermal time-scale introduced in equation (3.1).

It is clear that we must look for another source for the Sun's radiant energy, but before doing that we can deduce another very important result from equation (3.39). The total energy of a star can be defined by:

$$E = U + \Omega, \tag{3.42}$$

provided that there are no other sources of energy. If the star radiates energy into space, E must decrease. Combining equations (3.39) and (3.42) we have:

$$E = -U = \Omega/2. \tag{3.43}$$

The total energy of the star is negative and it is equal to half the gravitational energy or equivalently minus the thermal energy. We now see that a decrease in E leads to a decrease in Ω but an increase in U. Thus a star composed of a perfect gas, with no hidden energy supplies, contracts and heats up as it radiates energy. We thus have the rather paradoxical result that such a star finds it difficult to cool down; any attempt to lose energy causes the star to contract and to release energy at a rate which not only supplies the energy loss from the surface, but also heats up the material of the star. Although we have obtained this result for a fully ionized gas, it is true provided the ratio of specific heats γ exceeds 4/3. It is a very important result which we shall refer to when we consider the way in which a star evolves. We now see that if the Sun is a perfect gas it is impossible for cooling to be supplying its surface energy loss, regardless of the geological considerations.

We now return to a consideration of the source of the energy which the Sun and other stars radiate. If it is neither gravitational energy nor thermal energy, it seems that it must be released by the conversion of matter from one form to another. Moreover it must be capable of releasing at least 2×10^{-4} of the rest mass energy of the Sun (see page 59). This immediately rules out chemical reactions such as the combustion of coal, gas and oil which only release up to 5×10^{-10} of the rest mass energy. In fact the only known way in which quantities of energy as large as this can be released, by the change of matter from one form to another, is through nuclear reactions. These can be either fission reactions of heavy nuclei such as occur in the atomic bomb and nuclear reactors and which can release 5×10^{-4} of the rest mass energy or fusion reactions of light nuclei which occur in the hydrogen bomb and can release at most almost 1% of the rest mass energy. Not only are fusion reactions capable of a higher energy release, but also, as we have seen in Chapter 2, the light elements are the more abundant. It is now believed that nuclear fusion reactions are the source of the energy radiated during most phases of a star's evolution. This will be discussed more fully in the next chapter.

Relation between energy release and energy transport

Assuming that time variations are unimportant, we can immediately write down one equation relating the rate of energy release and the rate of energy transport. We suppose as before that the star is spherically symmetrical so that all energy is transported in the radial direction. Furthermore, the energy sources are distributed in such a way that the rate of energy release at radius r is $\varepsilon(r)$ W kg^{-1}. Suppose energy flows across a sphere of radius r at a rate L_r W. Then we can equate the difference between the energy crossing a sphere of radius $r+\delta r$ and a sphere of radius r to the energy released in the spherical shell (fig. 34).

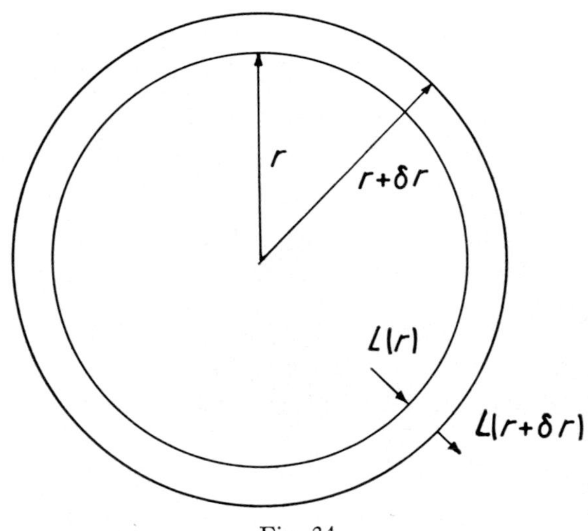

Fig. 34.

Thus:

$$L_{(r+\delta r)} - L_r = 4\pi r^2 \delta r \, \rho\varepsilon$$

or

$$\frac{dL}{dr} = 4\pi r^2 \rho\varepsilon. \tag{3.44}$$

In deriving this equation we have again neglected the change with time of the stellar properties. Thus, for example, we have not allowed for the possibility that some of the energy released in the shell is used to heat up or change the volume of the shell. This neglect of the time dependence will normally be justified if the energy sources at present being used are capable of supplying the star's radiation for a time long compared with its thermal time (3.41); in the notation of (3.1) this means:

$$t_n \gg t_{th}. \tag{3.45}$$

This is by no means always true and it is much more likely that time dependence has to be included in equation (3.44) than in equation (3.4)†.

Method of energy transport

We have obtained one further equation for the structure of a star, but only by introducing two more unknown quantities ε and L, so that it is clear that several more equations are still required. We must now consider the way in which energy is transported outwards in a star. There are three possible modes of energy transport; conduction, convection and radiation. Of these, there is no real distinction in principle between conduction and radiation, which both depend on the collision of energetic *particles* with less energetic *particles* resulting in an exchange of energy.

In the case of conduction, energy is mainly carried by electrons. The more energetic electrons from the hotter regions collide with electrons from the cooler regions and in the collision transfer energy. In the case of radiation the energy is carried by light quanta (photons). In most stars gas pressure is more important than radiation pressure (as we have shown for the Sun on page 59) and the same is true of the energy density. Thus:

$$\left.\begin{aligned} P_{\text{gas}} &= nkT, & \rho u_{\text{gas}} &= \frac{3}{2}nkT, \\ P_{\text{rad}} &= \frac{1}{3}aT^4, & \rho u_{\text{rad}} &= aT^4, \end{aligned}\right\} \tag{3.46}$$

where ρu is the energy per unit volume and the gas has been assumed to be monatomic. As the thermal energy of the electrons is much greater than the energy of the photons, it might be expected that thermal conduction would be the more important mechanism of energy transport of the two. However, there are two factors which determine the efficiency of transport of energy, the energy content, and the distance particles travel between collisions. The latter is known as the *mean free path.* If the mean free path is large, particles can get from a point where the temperature is high to one where it is significantly lower before colliding and transferring energy and a large transport of energy results. Although neither electrons nor photons can travel very far without col-

† It is possible to generalize equation (3.44) to take account of energy release which increases the thermal energy, or does work in changing the volume of an element of stellar material. The generalized equation is:

$$\frac{\mathrm{d}u}{\mathrm{d}t}+P\frac{\mathrm{d}v}{\mathrm{d}t} = \varepsilon-\frac{1}{4\pi r^2\rho}\frac{\partial L}{\partial r},$$

where the symbol $\mathrm{d}/\mathrm{d}t$ refers to the rate of change with time of the properties of a fixed element of material, u is the thermal energy per kg and v is the specific volume $(1/\rho)$. This equation has been used in the studies of stellar evolution, which will be described in the later chapters of this book.

lisions in typical conditions in stars, the photons have a considerably longer mean free path than the electrons and this more than offsets the greater total energy possessed by the electrons. In most stars the amount of energy carried by conduction is negligible compared to that carried by radiation.

Even so, it is possible to give an argument which suggests that the mean free path of photons in the Sun must be very small and that a photon must suffer many collisions in its journey from the interior to the surface of the Sun. Energy is released by nuclear reactions in the central regions of the Sun. If the photons released in these nuclear reactions travelled with the velocity of light to the surface of the Sun, they would escape from the Sun in a little over 2 s. In fact the energy released at the centre of the Sun slowly diffuses outwards. We have seen previously that the total thermal energy of the Sun could supply its rate of radiation for about 3×10^7 years. This gives us an estimate of how long it takes a photon to diffuse from the centre of the sun to the surface. When we observe the energy radiated at the solar surface we are usually seeing the results of nuclear reactions which occurred some tens of millions of years ago. Of course this is a rather crude discussion because photons do not retain their identity and what we have called collisions include absorption of radiation. However, it is certainly true that, if the Sun stopped releasing energy from nuclear reactions about ten million years ago, we should only be starting to notice the consequences now.

Whether conduction or radiation is being considered, the flux of energy per square metre per second (F) can be expressed in terms of the temperature gradient and a coefficient of thermal conductivity (λ). Thus:

$$\left.\begin{aligned} F_{\text{cond}} &= -\lambda_{\text{cond}} \mathrm{d}T/\mathrm{d}r, \\ F_{\text{rad}} &= -\lambda_{\text{rad}} \mathrm{d}T/\mathrm{d}r, \end{aligned}\right\} \tag{3.47}$$

in spherical symmetry, where the minus sign indicates that heat flows down the temperature gradient and, for example, $L_{\text{rad}} = 4\pi r^2 F_{\text{rad}}$. The thermal conductivity measures the readiness of heat to flow. The astronomer usually works in terms of a quantity which he calls the *opacity* of stellar material which measures the resistance of the material to the flow of heat. The opacity is defined by the relation:

$$\kappa = \frac{4acT^3}{3\rho\lambda}, \tag{3.48}$$

so that

$$F_{\text{rad}} = -\frac{4acT^3}{3\kappa_{\text{rad}}\rho} \frac{\mathrm{d}T}{\mathrm{d}r} \tag{3.49}$$

and

$$F_{\text{cond}} = -\frac{4acT^3}{3\kappa_{\text{cond}}\rho}\frac{dT}{dr}. \tag{3.50}$$

If all the transport of energy is by radiation and conduction, equations (3.49) and (3.50) can be combined to give:

$$L = 4\pi r^2(F_{\text{cond}}+F_{\text{rad}}) = -\frac{16\pi acr^2T^3}{3\kappa\rho}\frac{dT}{dr}$$

or

$$\frac{dT}{dr} = -\frac{3\kappa L\rho}{16\pi acr^2T^3}, \tag{3.51}$$

where

$$\frac{1}{\kappa} = \frac{1}{\kappa_{\text{rad}}}+\frac{1}{\kappa_{\text{cond}}}. \tag{3.52}$$

Equation (3.51) can also be rewritten:

$$\frac{dP_{\text{rad}}}{dr} = -\frac{\kappa L\rho}{4\pi cr^2}, \tag{3.53}$$

which is formally rather similar to equation (3.4):

$$\frac{dP}{dr} = -\frac{GM\rho}{r^2}.$$

Clearly the flow of energy by radiation and conduction can only be determined if an expression for κ is available. How such an expression can be obtained will be discussed in Chapter 4.

Energy transport by convection

Energy transport by conduction and radiation occurs whenever a temperature gradient is maintained in any body. The same is not true of convection which only occurs in liquids and gases, when the temperature gradient exceeds some critical value. Before we discuss convection in stars we will consider the much simpler problem of convection in a liquid which can be studied in a laboratory. Suppose (fig. 35) that

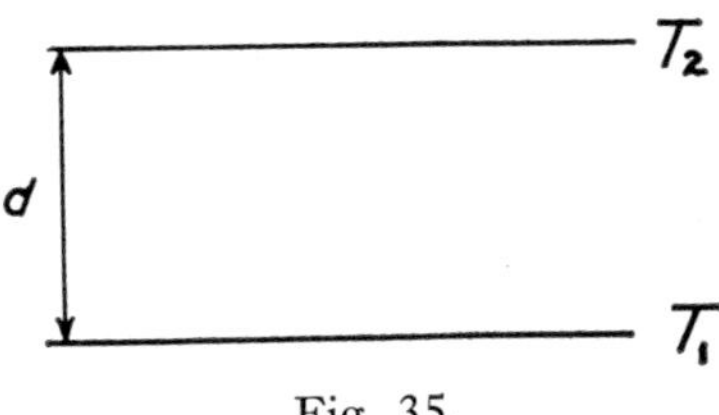

Fig. 35.

liquid is contained between two parallel surfaces which are maintained at different temperatures T_1 and T_2 with the lower surface being the hotter. For small temperature differences between the two walls, heat is carried only by conduction and the rate of flow of heat is given by the first of the two equations (3.47). If the temperature difference is increased, there is a critical stage when mass motions or convection occur in the liquid and the amount of energy transported increases sharply. At first the motions are fairly regular. The rising and falling elements of fluid form a fairly simple geometrical pattern; in special cases hexagonal convection cells are formed, with fluid rising at the centre of the hexagons and falling at the edges. If the temperature difference is increased still further, the regular patterns disappear and the motions become confused and irregular (turbulent).

Convection starts because the state of the fluid without motions is unstable. Suppose we ask what happens to a small element of fluid which moves upward from the lower surface. When this element has risen a small distance it is hotter than its surroundings and because of its coefficient of thermal expansion it is lighter than its surroundings. As it is lighter than its surroundings it will tend to rise further, but at the same time it will tend to conduct heat to its surroundings and cool down and the frictional force acting on the element will tend to slow it down. If the temperature gradient is high the buoyancy force will *win* and convective motions will start, whereas, if the temperature gradient is lower, the conduction of heat from the element and the resistance to motion offered by the surroundings are sufficient to prevent convection from starting.

Theoretical calculations have predicted that whether or not convective motions will occur should depend on the value of a dimensionless quantity known as the Rayleigh number. This is defined by:

$$R = g\alpha\beta d^4/\lambda\eta c_v, \tag{3.54}$$

where g is the acceleration due to gravity, α is the coefficient of thermal expansion, λ is the coefficient of thermal conductivity, η the viscosity and c_v the specific heat at constant volume of the liquid, d the depth of the layer of liquid and $\beta = |\mathrm{d}T/\mathrm{d}z|$. It is predicted that convection will occur if

$$R \gtrsim 1700, \tag{3.55}$$

and this has been confirmed experimentally with a variety of liquids. Once the Rayleigh number exceeds this critical value and convection is occurring, we are interested in how the amount of energy carried by convection is related to the Rayleigh number R. Here theory and observation are not in such good agreement. The amount of heat carried by convection when R is large can be measured, but at present there is no theory of fully developed convection which predicts the heat flow completely accurately.

Convection in stars is different from convection in a laboratory in several ways. In the first place there are no rigid walls maintained at well-defined temperatures; in fact, the size of a star and the values of the physical quantities within it depend on the way in which heat is transported. This means that we cannot first calculate the structure of the star and then ask how much energy is carried by convection. In the second place a star is composed of a highly compressible gas rather than an almost incompressible liquid. This means that a rising element of fluid in a star has a density which depends not only on its temperature, but also on its pressure, which is also the pressure of its surroundings. In a liquid convection would occur for any temperature gradient at all in a fluid heated below, if it were not for heat flow to the surroundings and the viscosity of the fluid. This is not true in a gas. In this case too, convection will occur if the rising element is lighter than its surroundings but this will depend on two things; the rate at which the element expands due to the decreasing pressure exerted on it and the rate at which the density of the surroundings decreases with height. Thus there must be a finite temperature gradient in a gas before convection will start, whatever the values of its thermal conductivity and viscosity. In the first instance we neglect the effect of heat losses and viscosity.

Condition for occurrence of convection

If a rising element loses no heat to its surroundings (this is called moving adiabatically) its pressure P and volume v obey the relation:

$$Pv^{\gamma} = \text{const}, \tag{3.56}$$

where, as before, γ is the ratio of the two principal specific heats. Equivalently (3.56) can be written:

$$P/\rho^{\gamma} = \text{const}. \tag{3.57}$$

Suppose an element rises a distance δz (fig. 36), starting with pressure P and density ρ and finishing with pressure $P-\delta P$ and density $\rho-\delta\rho$. Its

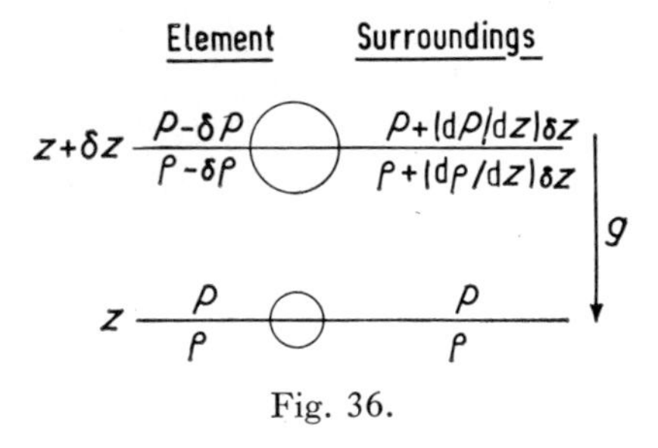

Fig. 36.

undisturbed surroundings at the upper level will have pressure, $P+(\mathrm{d}P/\mathrm{d}z)\delta z$, and density, $\rho+(\mathrm{d}\rho/\mathrm{d}z)\delta z$, where $\mathrm{d}P/\mathrm{d}z$ is negative as gravity acts downwards. From equation (3.57) it can be seen that

$$(P-\delta P)/(\rho-\delta\rho)^{\gamma} = P/\rho^{\gamma}. \tag{3.58}$$

In equation (3.58), if the distance δz and the changes δP, $\delta\rho$ are assumed small, $(\rho - \delta\rho)^\gamma$ can be replaced to sufficient accuracy by $\rho^\gamma - \gamma\rho^{\gamma-1}\delta\rho$ and equation (3.58) becomes:

$$\delta P = (\gamma P/\rho)\delta\rho. \tag{3.59}$$

As the element rises, it will remain at the same pressure as its surroundings so that

$$\delta P = (-\mathrm{d}P/\mathrm{d}z)\delta z. \tag{3.60}$$

Combining (3.59) and (3.60), we obtain:

$$\delta\rho = (\rho/\gamma P)\,(-\mathrm{d}P/\mathrm{d}z)\,\delta z. \tag{3.61}$$

The rising element will then be lighter than its surroundings and will continue to rise if

$$\rho - \delta\rho < \rho + (\mathrm{d}\rho/\mathrm{d}z)\delta z$$

or

$$(\rho/\gamma P)\,(\mathrm{d}P/\mathrm{d}z) < \mathrm{d}\rho/\mathrm{d}z, \tag{3.62}$$

where we have used (3.61) and divided the inequality by δz which is positive. We next divide both sides of (3·62) by $\mathrm{d}P/\mathrm{d}z$. As $\mathrm{d}P/\mathrm{d}z$ is negative, we must also change the sign of the inequality to give

$$\frac{P}{\rho}\frac{\mathrm{d}\rho}{\mathrm{d}P} < \frac{1}{\gamma}. \tag{3.63}$$

For a perfect gas in which radiation pressure is negligible, $P = \rho kT/m$, where m is the mean mass of the particles in the gas. Provided we are not considering a region in which ionization or dissociation is taking place and where m would vary with position,

$$\log P = \log \rho + \log T + \text{const}$$

and this can be differentiated to give

$$\frac{\mathrm{d}P}{P} = \frac{\mathrm{d}\rho}{\rho} + \frac{\mathrm{d}T}{T}. \tag{3.64}$$

Then equations (3.63) and (3.64) combine to give the condition for occurrence of convection as

$$\frac{P}{T}\frac{\mathrm{d}T}{\mathrm{d}P} > \frac{\gamma - 1}{\gamma}. \tag{3.65}$$

In deriving this condition we have not considered the spherical geometry of a star. This is not, however, a defect as the criterion for the onset of convection only depends on conditions in the immediate neighbourhood of the element concerned and in a small enough region it is impossible to distinguish between a plane and spherical system†.

† This corresponds to the use of plane geometry on small regions of the Earth's surface.

(3.65) can also be written:

$$\left|\frac{\mathrm{d}T}{\mathrm{d}z}\right| > \left(\frac{\gamma-1}{\gamma}\right)\frac{T}{P}\left|\frac{\mathrm{d}P}{\mathrm{d}z}\right|, \tag{3.66}$$

where the modulus signs are used because both $\mathrm{d}P/\mathrm{d}z$ and $\mathrm{d}T/\mathrm{d}z$ are negative, and it can be seen that convection will occur if $|\mathrm{d}T/\mathrm{d}z|$ exceeds a certain multiple of $|\mathrm{d}P/\mathrm{d}z|$. As in the case of a liquid a somewhat higher temperature gradient than this will be required before convection occurs because of effects of heat flow from the element and viscosity. However, this correction to the critical temperature gradient is usually small compared to the gradient itself and we shall ignore it in what follows.

Once again we have determined when convection is likely to occur, but we also need to know how much energy will be carried by convection when criterion (3.65) is thoroughly violated. It is not possible to do experiments which mimic the conditions inside stars and at present there is not a generally accepted theory which calculates how much energy will be carried by fully developed convection. Fortunately, as we shall see later, there are occasions when we can manage without this knowledge, but *the lack of a good theory of convection is one of the worst defects in our present studies of stellar structure and evolution.*

The structure of stars

If for the moment we assume that convection does not occur, we have four differential equations governing the structure of stars:

$$\frac{\mathrm{d}P}{\mathrm{d}r} = -\frac{GM\rho}{r^2}. \tag{3.4}$$

$$\frac{\mathrm{d}M}{\mathrm{d}r} = 4\pi r^2\rho, \tag{3.5}$$

$$\frac{\mathrm{d}L}{\mathrm{d}r} = 4\pi r^2\rho\varepsilon \tag{3.44}$$

and

$$\frac{\mathrm{d}T}{\mathrm{d}r} = -\frac{3\kappa L\rho}{16\pi acr^2T^3}. \tag{3.51}$$

Included in these equations are three quantities, P, κ and ε, which we shall consider further in the next chapter. We can say now that, if the star is in a steady state and in a state close to thermal equilibrium, all of these quantities should depend on the density, temperature and chemical composition of the star where, of course, all of these will in general be functions of the radius r. We have already written down in equations (3.25) and (3.27):

$$P_{\text{gas}} = nkT, \quad P_{\text{rad}} = \frac{1}{3}aT^4,$$

possible expressions for the pressure and it is a problem of basic physics to tell us what are the pressure, opacity and rate of energy generation of a medium for given conditions of density and temperature. This we shall discuss further in the next chapter, but at present we can assume that such expressions can be obtained and write:

$$P = P(\rho,\ T,\ \text{composition}), \tag{3.67}$$

$$\kappa = \kappa(\rho,\ T,\ \text{composition}) \tag{3.68}$$

and

$$\varepsilon = \varepsilon(\rho,\ T,\ \text{composition}). \tag{3.69}$$

Given the chemical composition of the star we now have seven equations for the seven unknowns P, ρ, T, M, L, κ and ε as functions of r.

Calculations of the structure of a star now involve obtaining expressions for P, κ and ε and then the solution of the four differential equations (3.4), (3.5), (3.44) and (3.51). In general such a solution can only be obtained with the aid of a computer and, if the best possible expressions are used for P, κ and ε, quite a large computer is required. Before the equations can be solved we must consider what are known as boundary conditions. If we consider a single first-order differential equation of the form

$$\frac{\mathrm{d}y}{\mathrm{d}x} = f(x,\ y),$$

it will not have a unique solution but will usually have a solution containing one arbitrary constant. The solution can be made precise if we know in advance the value that y must take at one specific value of x. In many physical problems we know the value of y at one of the boundaries of the system and this is then known as a boundary condition. In the problem which we are studying in this chapter, we have not one but four first-order differential equations and we can expect to be able to satisfy four boundary conditions. Indeed, unless we have four boundary conditions, it will be impossible to determine the structure of the star uniquely.

Two of the boundary conditions are at the centre of the star and two at the surface. The two at the centre are quite obvious. The mass and the luminosity are both quantities which must increase outwards from zero at the centre of the star. Thus:

$$M = 0, \quad L = 0 \quad \text{at } r = 0. \tag{3.70}$$

The surface boundary conditions are not so clear, but simple approximations to these conditions are often used. If stars were truly isolated bodies, we would expect the density and the pressure to fall to zero at the surface. In fact there is no sharp edge to a star, but the density of the Sun near to the visible surface is estimated to be about 10^{-4} kg m^{-3} which is extremely small compared to the value of the mean density

$1{\cdot}4 \times 10^3$ kg m^{-3} given by equation (3.34). Similarly the surface temperatures of stars are very much smaller than their mean temperatures. In the case of the Sun, its surface temperature of 5780 K can be compared with the typical temperature 2×10^6 K predicted by equation (3.33). Since the surface temperature and density are both very small compared to typical values of these quantities, it seems plausible that the solution of the equations of stellar structure throughout most of the interior of a star will not be seriously affected if the true surface boundary conditions are replaced by the assumption that the density and temperature vanish at the surface. Thus we take:

$$\rho = 0, \quad T = 0 \quad \text{at} \quad r = r_s. \tag{3.71}$$

Clearly the use of these approximate boundary conditions will not allow us to obtain detailed information about the properties of the outer layers of a star. However, in many cases the radius and luminosity of a star obtained using the correct boundary conditions differ negligibly from those obtained using the approximate boundary conditions. There are also some occasions when the boundary conditions (3.71) give a poor representation not only of the surface layers, but also of the whole internal structure of the star. Such a case will be discussed in Chapter 5. It might be thought that the use of a boundary condition such as (3.71) makes impossible the determination of the surface temperature of a star, which is one of the quantities which our theory should predict. However, once we know the luminosity and radius of a star we can calculate the effective temperature, which is the temperature of a black body with the same radius and luminosity as the star. Thus, from Chapter 2:

$$L_s = \pi a c r_s^{\,2} T_e^{\,4}. \tag{2.7}$$

This effective temperature can be compared with observational estimates of surface temperature. The uncertainty involved in converting theoretical effective temperature into observed colour indices has already been discussed in Chapter 2.

Use of mass as independent variable

In the introduction to this chapter, it has been explained that the theoretical astrophysicist does not usually try to calculate the properties of a particular star. Instead he studies a wide variety of possible stars. These possible stars can be defined by their *mass* and *chemical composition*: the amount of matter they contain and what kind of matter it is. From the theoretical point of view the mass of a star is to be regarded as something to be chosen before the equations of stellar structure are solved, while the radius is something to be determined from these calculations. For this reason it is often inconvenient that the boundary conditions (3.71) have to be applied at a radius which is not known in advance.

To avoid this difficulty, it is often useful to write the equations in terms of M as the independent variable instead of r. Thus division of equations (3.4), (3.44) and (3.51) by (3.5) and the inversion of (3.5) itself, give:

$$\frac{dP}{dM} = -\frac{GM}{4\pi r^4}, \tag{3.72}$$

$$\frac{dr}{dM} = \frac{1}{4\pi r^2 \rho}, \tag{3.73}$$

$$\frac{dL}{dM} = \varepsilon, \tag{3.74}$$

$$\frac{dT}{dM} = -\frac{3\kappa L}{64\pi^2 acr^4 T^3}. \tag{3.75}$$

Similarly the boundary conditions (3.70), (3.71) can be written:

$$r = 0, \quad L = 0 \quad \text{at} \quad M = 0 \tag{3.76}$$

and

$$\rho = 0, \quad T = 0 \quad \text{at} \quad M = M_s, \tag{3.77}$$

where now M_s is assumed known for any particular calculation.

We now specify the mass and chemical composition of a star and have a well-defined problem to solve equations (3.72)–(3·75) with the subsidiary relations (3·67)–(3.69) and the boundary conditions (3·76) and (3.77). In the astronomical literature there is to be found a *theorem* known as the Vogt–Russell theorem which states that, if the mass and chemical composition of the star are specified, this set of equations has only one solution and the structure of the star is uniquely determined. However, no rigorous proof of this *theorem* was given and it is now known that in some rare circumstances two different solutions of this set of equations may be possible. In that case which solution is relevant must be determined by the past history of the star and by a set of equations in which time dependence occurs. In this book we shall assume that the Vogt–Russell theorem is valid.

Stellar evolution

In this chapter we have obtained a set of equations which determine the structure of a star of given mass and chemical composition and, as these equations contain no time derivatives of the physical quantities, they cannot tell us how the properties of a star change with time. But stars do evolve. They are continually radiating energy into space and thereby losing mass and, as this energy is released by nuclear reactions in the interior of the star, the chemical composition is also changing. This mass loss is slight and cannot exceed 1% of the star's mass in its entire life (although a star might lose mass directly through other causes and this certainly happens in the explosions of novae and supernovae) and this can safely be ignored, but the change in chemical composition is crucial in determining how the properties of star change as it evolves.

The evolution of a star can only be studied by using equations in which the time derivatives of the physical quantities are included. We have previously argued that the time derivative can be omitted from equation (3.4) provided that evolution is occurring slowly compared to the dynamical time-scale:

$$t_d = (2r^3/GM)^{1/2}, \tag{3.11}$$

and that will certainly be true for normal stellar evolution. We have also suggested that time dependence can be omitted from equation (3.44) provided that the series of nuclear reactions at present occurring can supply the star's radiation for a time greatly in excess of the thermal time. That is:

$$t_n \gg t_{th}. \tag{3.45}$$

As we shall see in later chapters of this book there are many stages in stellar evolution when inequality (3.45) is not true and in these stages the evolution can only be calculated by using a more accurate version of equation (3.44).

However, there are occasions in stellar evolution when inequality (3.45) is satisfied and the set of equations (3.72)–(3.75) is a perfectly adequate set of equations. How then can the evolution of the star be studied? The nuclear reactions which are supplying the energy the star is radiating are changing its chemical composition and we can therefore write down equations describing how this change of composition is occurring.

If there are no bulk motions in the interior of a star, such as would occur if convection were the main mechanism of energy transport, any changes of chemical composition are localized in the element of material in which they are brought about by nuclear reactions. This means that, even if the star started its life with a homogeneous chemical composition, the subsequent composition must be regarded as a function of the mass M. In this simplest case, in which no bulk motions occur, the set of equations must be supplemented by equations describing the rate of change with time of the abundances of the different chemical elements. These equations may be schematically written:

$$\frac{\partial}{\partial t}(\text{composition})_M = f(\rho, T, \text{composition}), \tag{3.78}$$

where the right-hand side has the form shown because the rate at which nuclear reactions occur depends on ρ, T and composition. In the case when hydrogen is being converted into helium we would have two equations of the form of (3.78). One would describe the decrease in the hydrogen content and the other the increase in the helium content.

Method of solution of the equations

Now we have a time dependent equation (3.78), the derivatives d/dM which occur in equations (3·72)–(3.75) should strictly be replaced by the partial derivatives $(\partial/\partial M)_t$ but this does not really matter as there are no equations in which derivatives with respect to both M and t occur and the time and mass dependence of the problem are completely separated. To show what we mean by this we will consider how the evolution of a star could be studied. We first suppose that we know the chemical composition of a star as a function of M at some time t_0; we will not enquire at the moment how we can know this. If the total mass M_s and the chemical composition as a function of M are known, equations (3.72)–(3.75), (3.67)–(3.69) can be solved with the boundary conditions (3.76) (3.77) to determine the complete internal structure of the star at time t_0.

The evaluation of the expression (3.69) for the rate of energy release, ε, has involved a discussion of the rate at which different nuclear reactions are occurring and thus how the chemical composition of the star is changing. The new chemical composition of the star as a function of M at the time $t_0 + \delta t$ can now be obtained from the (schematic) equation:

$$(\text{composition})_{M, t_0+\delta t} = (\text{composition})_{M, t_0} + \frac{\partial}{\partial t}(\text{composition})_M \delta t, \tag{3.79}$$

where, for example, $(\text{composition})_{M, t_0}$ denotes the composition at mass M at time t_0. Use of equations (3.79) at all points in the star gives the chemical composition as a function of M at time $t_0 + \delta t$. With this modified chemical composition equations (3.72)–(3.75), (3.67)–(3.69) can again be solved to find the modified structure of the star so that we know how the star's properties have changed between times t_0 and $t_0 + \delta t$. This process can be repeated and the life history of the star can be described.

This process will break down if at any time the properties of the star are found to be changing so rapidly with time that time dependent terms in equations (3.72)–(3.75) cannot be regarded as unimportant. In that case these equations must be modified. The problem then becomes more difficult mathematically because there are now equations containing derivatives with respect to both M and t and the mass and time dependence of the problem no longer separates as in our discussion above. The equations can, however, be solved and the results of such calculations will be described in later chapters of the book.

Influence of convection

The above discussion is also over-simplified because in most stars there is a region in which a significant amount of energy is carried by convection. Suppose we have calculated the structure of a star of given mass and chemical composition assuming that convection does not occur.

We can now check whether that assumption is valid. At all points in the star we have values for P, $\mathrm{d}P/\mathrm{d}M$, ρ and $\mathrm{d}\rho/\mathrm{d}M$ and we can hence calculate $(P\mathrm{d}\rho/\mathrm{d}M)/(\rho\mathrm{d}P/\mathrm{d}M)$ and compare its value with $1/\gamma$ to see whether or not criterion (3.63) is satisfied anywhere. If it exceeds $1/\gamma$ at all points in the star, convection is indeed absent and the structure of the star has been correctly calculated but, if it is less than $1/\gamma$ anywhere, convection must be occurring and the whole solution of the equations must be reconsidered.

The modification of the equations is as follows. Instead of equation (3.75) we must use the equation:

$$L_{\mathrm{rad}} = -\frac{64\pi^2 acr^4T^3}{3\kappa}\frac{\mathrm{d}T}{\mathrm{d}M}, \tag{3.80}$$

which says that whether or not convection is occurring the amount of energy carried by radiation (which also includes conduction as usual) is determined by the temperature gradient. We can then write:

$$L = L_{\mathrm{rad}} + L_{\mathrm{conv}} \tag{3.81}$$

and we require an expression for the amount of energy carried by convection, L_{conv}. We write this final equation schematically as:

$$L_{\mathrm{conv}} = ?, \tag{3.82}$$

where we indicate that there is no reliable theory which calculates the amount of energy carried by convection and we do not write $L_{\mathrm{conv}}(\rho, T,$ composition), because it is not clear that the amount of energy carried by convection across any radius r depends only on conditions at that radius. Such a form will be valid if typical convective elements only travel a short distance, but not if many of them have come from regions with very different physical conditions. In the case of convection in a liquid it is known that the amount of energy carried by convection depends on the depth of the layer of liquid in which convection is occurring.

If we had an expression for the right-hand side of equation (3.82), the equations of stellar structure would be modified by the replacement of equation (3.75) by (3.80) to (3.82); of course the expression of (3.82) would ensure that no energy was carried by convection if $P\mathrm{d}T/T\mathrm{d}P < (\gamma-1)/\gamma$. It would then still be a reasonably straightforward matter to solve for the structure of a star of given mass and chemical composition. The study of stellar evolution is more difficult because the changes of chemical composition are now not necessarily localized where they occur.

The problem is still reasonably simple if convection only occurs in a region in which there are no nuclear reactions or if the region of nuclear energy release is entirely contained within a convective region. In the former case, equation (3.78) is still valid. In the latter case the chemical composition of the convective region probably remains essentially uniform as convection currents keep the region well mixed. In the

Sun, for example, significant changes of chemical composition cannot at present be occurring in less than about 10^9 years. Convection currents would only have to have a speed of 10^{-7} m s^{-1} to travel across the Sun's radius several times in that time and they would have to be very much faster than that if they were to carry much energy.

We have said that convection will occur if criterion (3.65) is satisfied. Broadly speaking there are two ways in which this can happen. Either, for a gas with a normal value of γ, the temperature gradient required to carry all of the energy by radiation can become large or there may be a region in which the ratio of specific heats becomes close to unity and the criterion can be satisfied for an ordinary value of the temperature gradient. If a large amount of energy is released in a small volume at the centre of a star, it may require a large temperature gradient to carry the energy away. This means that convection may occur in regions in which nuclear energy is released near the centre of stars and such regions are known as convective cores. Specific heats at constant pressure and constant volume approach equality when a change of state is occurring and most of the energy supplied in an attempt to raise the temperature is used up in supplying the latent heat for the change of state. If the surface temperature of a star is low enough for the atoms at the surface to be predominantly neutral, there may be a zone just below the surface in which the abundant elements are being ionized and in which the ratio of specific heats is close to unity. In such a case a star can have an outer convective layer.

Convection in stellar interiors

Although no really adequate expression is available for the right-hand side of equation (3.82), there are fortunately occasions on which, even though convection occurs, it is possible to avoid using equation (3.82). In the deep interior of a star in which there is a convective core, it appears that only a very slight increase of the temperature gradient over the adiabatic value defined by

$$\left(\frac{P}{T}\frac{\mathrm{d}T}{\mathrm{d}P}\right)_{\mathrm{ad}} = \frac{\gamma-1}{\gamma} \tag{3.83}$$

occurs before convection is capable of carrying *all* of the energy which is required.

An approximate estimate of how much energy can be carried by convection can be made as follows. Heat is convected by rising elements which are hotter than their surroundings and by falling elements which are cooler than their surroundings. Suppose that each type of element has a temperature which differs by δT from that of the surroundings. As a rising element is in pressure balance with its surroundings, it has an energy content per kilogramme which is $c_{\mathrm{p}}\delta T$ greater than the energy content of a kilogramme of the surrounding medium, where c_{p} is the specific heat at constant pressure. If we consider the stellar material to be a monatomic perfect gas c_P is $5k/2m$, where m is the average mass of the particles in the gas. The falling elements have a similar energy deficit. Suppose that a fraction $\alpha(\leqslant 1)$ of the material is in the rising and falling columns and suppose that they are both moving with a speed v m s^{-1}. Then the rate at which excess energy is carried across a sphere of radius r is:

surface area of sphere × rate of transport of mass × excess energy per unit mass

$$= 4\pi r^2 . \alpha\rho v .\ 5k\delta T/2m$$
$$= 10\pi r^2 \alpha\rho v k \delta T/m. \quad (3.84)$$

Fairly near the centre of the Sun we can consider a sphere of radius 10^8 m and the density of the material is about 5×10^4 kg m^{-3}†. With k equal to $1{\cdot}4 \times 10^{-23}$ J K^{-1} and m taken equal to 8×10^{-28} kg (the value appropriate to fully ionized hydrogen):

$$L_{\text{conv}} \simeq 2{\cdot}5 \times 10^{26}\ \alpha v \delta T \text{ W}. \quad (3.85)$$

At no point in the Sun does the luminosity exceed its surface value of 4×10^{26} W. Thus it can be seen from (3.85) that, provided a reasonable fraction of the material is taking part in the convection, a velocity of a few metres a second and a temperature difference of a few degrees suffice to carry all of the Sun's energy. As the actual temperature in the solar interior is about 10^7 K and the velocity of sound, which is essentially the random velocity of the particles, is about 4×10^5 m s^{-1}, this suggests that very mild convection indeed would suffice to carry all of the Sun's energy. As the temperature excess and velocity of the rising elements is determined by the difference between the actual temperature gradient and the adiabatic gradient‡, this suggests that the actual gradient is not greatly in excess of the adiabatic gradient. To a reasonable degree of accuracy we can assume that the temperature gradient has exactly the adiabatic value in a convective core. Thus we take:

$$\frac{P}{T}\frac{\mathrm{d}T}{\mathrm{d}P} = \frac{\gamma - 1}{\gamma}. \quad (3.86)$$

Note although we have used solar values in this discussion, the same result is valid in other stars. Note also that the Sun may not have a convective core; all we have shown here is that, if it does have a convective core, equation (3.86) will be almost exactly true in it.

When we have a convective region in which equation (3.86) is essentially true, we can forget about equations (3.81) and (3.82) and we

† This value comes from solutions of the equations of stellar structure but the argument could be made with mean density $1{\cdot}4 \times 10^3$ kg m^{-3}.

‡ In the most commonly used theory, the energy carried by convection depends on the (3/2)th power of this difference.

can replace equation (3.80) by equation (3.86). Thus, in a convective core, the four differential equations

$$\frac{dP}{dM} = -\frac{GM}{4\pi r^4}, \tag{3.72}$$

$$\frac{dr}{dM} = \frac{1}{4\pi r^2 \rho}, \tag{3.73}$$

$$\frac{dL}{dM} = \varepsilon \tag{3.74}$$

and

$$\frac{P}{T}\frac{dT}{dP} = \frac{\gamma - 1}{\gamma} \tag{3.86}$$

must be solved together with equations for ε and P. Equation (3.80) for the radiative flux:

$$L_{rad} = -\frac{64\pi^2 acr^4 T^3}{3\kappa}\frac{dT}{dM}, \tag{3.80}$$

is of course still true and once the other equations have been solved L_{rad} can be calculated. This can then be compared with L calculated from (3.74) and the difference between the two gives the value of L_{conv}. If convection is occurring, L_{conv} must be positive and if at any time it is found to be negative, this is a signal that the temperature gradient given by (3.86) is more than capable of carrying all of the energy by radiation and that convection will not in fact occur. In an actual star there will possibly be regions in which convective occurs and other regions in which it does not. In solving the equations of stellar structure, the equations appropriate to a convection region must be *switched on* whenever the temperature gradient reaches the adiabatic value and these convective equations must be *switched off* whenever radiation is capable of carrying all of the energy with a temperature gradient lower than the adiabatic value.

Convection near stellar surface

Although, in the central regions of stars, equation (3.86) is valid if convection occurs, the same is not true near the surface. Thus near the surface of the Sun, where r is 7×10^8 m and $\rho \simeq 10^{-3}$ kg m^{-3}.

$$L_{conv} \simeq 2{\cdot}5 \times 10^{20}\, \alpha v \delta T \text{ W}. \tag{3.87}$$

In this region the temperature is about 10^4 K and the velocity of sound 10^4 m s^{-1}. It is clear that, in order for the convective luminosity to be comparable with the total energy transport of 4×10^{26} W, the velocity of rising and falling elements must be comparable with the velocity of sound and the temperature differences must be comparable with the actual temperature. Such velocities and temperature differences can only be driven by a temperature gradient substantially in excess of the adiabatic value and in such a case an expression for the amount of energy carried by convection (3.82) is required. It is in such low density surface regions that the lack of a good theory of convection can be serious.

Summary of Chapter 3

In this chapter the equations of stellar structure have been formulated. Stars are held together by the force of gravitation and the gravitational force on unit volume is resisted by the pressure gradient of the stellar material. Because the properties of most stars are slowly varying, these forces must be almost exactly in balance and from this it is possible to deduce that the temperatures inside stars are higher than 10^6 K and that, despite stellar densities being as high as solids, the stellar material is gaseous. It is, in fact, a gas of ions and electrons rather than atoms and molecules. If it is a perfect gas, it follows from the Virial Theorem that a star becomes hotter as it loses energy.

The luminosity of the Sun has not changed significantly in the past 10^9 years and only energy released by nuclear fusion reactions can have supplied all of this energy. Nuclear energy is released at the hottest regions in a star near its centre and it must be carried to the surface by conduction, convection or radiation. The transport of energy by conduction and radiation depends on reasonably well-understood physical processes and it appears that conduction is unimportant in most stars. The condition for the onset of convection is clear, but there is at present no good theory which predicts how much energy is carried by fully developed convection. Fortunately, in many cases, convection can be shown to be so efficient that it will carry all of the energy required and then a detailed theory of convection is not needed.

The calculation of the structure of a star requires the specification of its mass and chemical composition and the solution of four differential equations with two boundary conditions at both the centre and the surface. In many stages of a star's evolution, all time derivatives are small and can be omitted from the equations. It is then relatively simple to study the slow evolution of a star. The structure of the star can first be calculated at a given time. Nuclear reactions then cause a gradual change in its chemical composition and the structure can be recalculated a short time later with the revised chemical composition. This procedure can then be repeated.

CHAPTER 4

the physics of stellar interiors

Introduction

In the last chapter we have derived the equations of stellar structure. In these equations there are three quantities, the pressure, P, the energy release per kilogramme per second, ε, and the opacity coefficient κ which depend only on the density, temperature and chemical composition of the stellar material. In Chapter 3 we did not discuss how P, ε and κ depend on these quantities, and the present chapter is concerned with discussing how P, ε and κ may be calculated if the density, temperature and chemical composition are known. To calculate ε a considerable knowledge of nuclear physics is required and a similar knowledge of atomic physics is required for the determination of κ. All three quantities depend on the thermodynamic state of the stellar material. Once ρ, T and the chemical composition are known, the calculation of P, ε and κ is pure physics and no further astronomical concepts are required and it is for this reason that this chapter is called *The Physics of Stellar Interiors*.

Because of the great complexity of the problems involved we are only able to describe the basic processes determining P, ε and κ and are not able to give detailed calculations. In the first place, we consider the law of energy release.

Energy release from nuclear reactions

As mentioned in the last chapter, it is now believed that most of the energy radiated by stars has been released by nuclear reactions in the stellar interior. We first consider why it is that energy can be released by nuclear reactions and how we determine which nuclear reactions will release energy.

Atomic nuclei are composed of protons and neutrons, which together are referred to as nucleons. The total mass of a nucleus is less than the mass of its constituent nucleons. That means that there is a decrease in mass if a compound nucleus is formed from nucleons and by the Einstein mass–energy relation, $E = mc^2$, this lost mass is released as energy. This energy is known as the binding energy of the compound nucleus and this can be calculated when the mass difference between the compound nucleus and its constituent nucleons is known. Thus, if a nucleus is composed of Z protons and N neutrons, its binding energy $Q(Z, N)$ is:

$$Q(Z,N) \equiv [Zm_{\mathrm{p}} + Nm_{\mathrm{n}} - m(Z, N)]c^2, \tag{4.1}$$

where m_p is the proton mass, m_n the neutron mass and $m(Z,N)$ the mass of the compound nucleus.

A more significant quantity for our present discussion than the total binding energy of a nucleus is the binding energy per nucleon $Q/(Z+N)$. This is proportional to the fractional loss of mass when the compound nucleus is formed. If $Q(Z,N)/(Z+N)$ is plotted against $A(\equiv Z+N)$ for a large number of nuclei, the resulting diagram has the general character shown in fig. 37. The actual curve is very irregular; in particular for a given value of A there may be several isobars (nuclei which

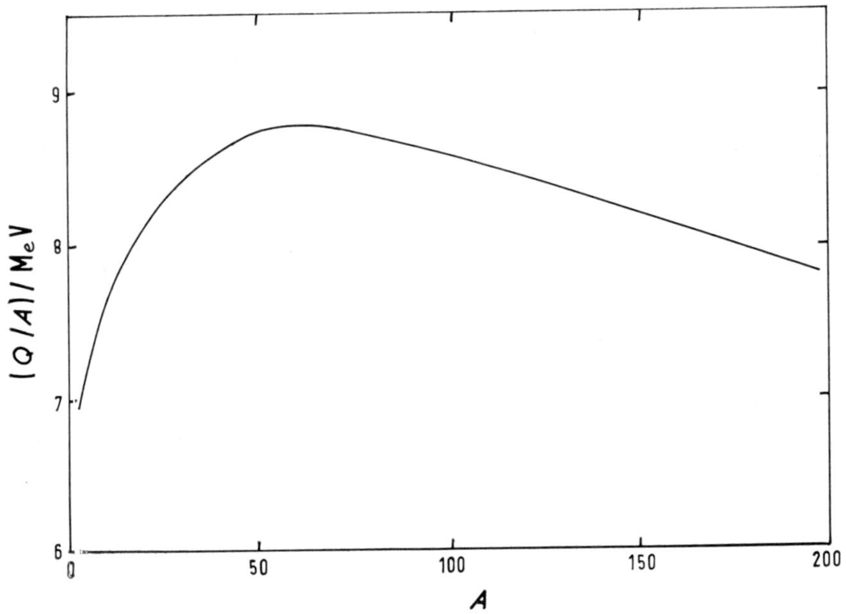

Fig. 37. The binding energy per nucleon as a function of atomic mass number; Q is measured in MeV.

have the same number of nucleons, but a different division between protons and neutrons) which have different values of Q. The general property of the curve is that the binding energy per nucleon rises rapidly with the initial increase in nucleon number, there is a broad maximum for A values between 50 and 60 (that is for nuclei in the neighbourhood of iron in the periodic table) and then there is a gradual decline for nuclei with higher values of A. The nuclei in the neighbourhood of iron are the most strongly bound nuclei. They have the greatest fractional loss of mass when formed out of individual protons and neutrons.

From fig. 37 the possibility of nuclear fusion and fission reactions releasing energy can be deduced. Thus consider first the process by which light nuclei combine (fuse) to form heavier nuclei. If the two nuclei and their compound all lie to the left of the maximum in fig. 37, the compound nucleus has a larger binding energy per nucleon than the original nuclei and, as the total number of nucleons has not been changed, the nuclear reaction must release energy. An example of such an energy releasing fusion reaction is the combination of a helium nucleus and a carbon nucleus to produce an oxygen nucleus:

$$^{4}\text{He} + {}^{12}\text{C} \rightarrow {}^{16}\text{O} + 7{\cdot}2\ \text{MeV}. \tag{4.2}$$

If a succession of fusion reactions takes place in which no more than a few nucleons are added at any time, further energy release from fusion reactions becomes impossible when the nucleus is in the iron region.

Fission reactions, in which a heavy nucleus splits into two or more fragments, will also clearly release energy if all of the nuclei involved are to the right of the maximum in fig. 37. In this case also the newly formed nuclei have a larger binding energy per nucleon than the initial nucleus so that it is clear that energy has been released by the reaction†. In the case of the heaviest nuclei these fission reactions occur spontaneously as in the fission of ^{235}U.

Fission reactions provide the energy release in the atomic bomb and in nuclear reactors. The fusion of light nuclei provides the energy of the hydrogen bomb. In addition, for the past twenty years scientists in many countries have been trying to produce nuclear fusion reactions between the heavy isotopes of hydrogen, deuterium and tritium, under controlled conditions in the laboratory. The problem of producing controlled thermonuclear reactions appears to be very difficult because the material must be raised to a temperature in excess of 10^8 K and kept at that temperature long enough to produce a useful yield of energy. At the same time the hot reacting material must be insulated from the walls of the containing vessel. If these difficulties are eventually overcome, the world's useable energy resources will be vastly increased.

The energy release from fusion reactions converting hydrogen into the most abundant isotopes of helium (^{4}He) and iron (^{56}Fe) is, in J kg^{-1}.

$$\begin{aligned} \text{H} &\rightarrow {}^{4}\text{He}, \quad 6{\cdot}3 \times 10^{14}, \\ \text{H} &\rightarrow {}^{56}\text{Fe}, \quad 7{\cdot}6 \times 10^{14}. \end{aligned} \tag{4.3}$$

The latter is about the maximum energy release which can possibly be obtained from nuclear fusion reactions and, since the rest mass energy of 1 kg is 9×10^{16} J, this maximum energy release is just under 1% of the rest mass energy as stated in Chapter 3. The binding energy per nucleon of very heavy nuclei, although less than that of iron, is still quite large. This means that the maximum possible energy release per kg from fission reactions is much less than that from fusion reactions.

† In actually occurring fission reactions some of the fragments usually lie to the left of the maximum in the binding energy curve and a more careful discussion is required to demonstrate the energy release.

Thus, if there were comparable amounts of heavy and light elements in stars, we would expect fusion reactions to be a more important source of energy than fission reactions. In fact, as we have seen in fig. 21 of Chapter 2, the very heavy nuclei do not appear to be very abundant in nature and it is thus believed that nuclear fusion reactions are by far the main source of the energy radiated by stars.

The four forces of physics

From fig. 37 we have seen that nuclear fusion reactions are energetically possible, but we must now discuss the conditions under which they will actually occur and decide whether these conditions will be found in stars. For example, it is obviously not true that hydrogen changes spontaneously into iron under all conditions. In stellar interiors the material is highly ionized and we are interested in reactions between bare nuclei. Nuclei interact through the four basic interactions of physics: electromagnetic, gravitational, strong nuclear and weak nuclear. Of these the two most important in the present subject are the electromagnetic and strong nuclear interactions. Gravitational forces are vitally important for the structure of whole stars, but are completely negligible in interactions between individual particles. A measure of the weakness of the gravitational interaction is the ratio of the gravitational force between a proton and an electron to the electrostatic force between the same particles. This is:

$$4\pi\varepsilon_0 G m_p m_e / e^2 \simeq 4 \times 10^{-40}, \tag{4.4}$$

where m_e and e are the mass and charge of the electron respectively.

As the electrostatic and gravitational interactions are both inverse square forces such a comparison is easily made. The strong and weak nuclear interactions do not share this property as they are both short range interactions which means that they are only important when particles are very close together. In the region where they are important, the strong nuclear interaction is much stronger than the electrostatic interaction, while the weak nuclear interaction, although weaker than the electrostatic interaction, is very much stronger than gravitational forces. The particular importance of the weak interaction is that, if it did not exist, many unstable elementary particles, such as the neutron, would be stable. Thus neutron decay

$$\mathrm{n} \rightarrow \mathrm{p} + \mathrm{e}^- + \bar{\nu} \tag{4.5}$$

in which a neutron (n) is converted into a proton (p), an electron (e^-) and an anti-neutrino† ($\bar{\nu}$) would not occur if it were not for the existence

† Every elementary particle possesses an anti-particle which has equal mass but equal and opposite values for other basic properties such as electric charge. For example, the positron is the anti-particle of the electron. There are two neutrinos, one of which is arbitrarily called the neutrino and the other the anti-neutrino. When a positron is emitted in β-decay it is accompanied by a neutrino while an electron is accompanied by an anti-neutrino.

of the weak nuclear interaction. We shall see that a similar reaction converting a proton into a neutron plays an important part in the conversion of hydrogen into helium as in that conversion two protons must be changed into neutrons.

We first consider only the electromagnetic and strong nuclear interactions between nuclei. The strong nuclear interaction has a very short range and two nucleons only affect one another through it if they are separated by less than about 10^{-15} m. At this separation and rather less, the interaction is attractive and it is the force which holds nuclei together and which causes the nuclear reactions which occur in stars. Thus if two nuclei come close enough to be influenced by it, they are drawn together and may form a compound nucleus.

The electrostatic force between two positively charged nuclei is repulsive and, unlike the strong nuclear interaction it has a long range, only falling off with distance, r, as $1/r^2$. The existence of stable nuclei depends on the balance between these two forces. Because all of the protons in a nucleus repel all of the other protons through their electrostatic interaction and because, in contrast, only the nearby protons and neutrons strongly attract one another through the nuclear force, in heavy nuclei the proportion of nucleons in the form of protons decreases. If there were more protons, the repulsive electrostatic force would be too strong to allow the nucleus to exist. In the case of ^{238}U for example there are 146 neutrons and only 92 protons.

Occurrence of nuclear reactions

If we consider not the structure of a given nucleus but the possibility that a nuclear interaction between two nuclei can lead to a nuclear transmutation, we see that the repulsive electromagnetic interaction between two positively charged nuclei tends to prevent them from approaching close enough for the strong nuclear interaction to become significant. Nuclear reactions can occur if particles approach close enough for this to happen despite the repulsive effect of the electric charges and this occurs if the relative velocity of the two particles is high enough.

This is illustrated in fig. 38. A co-ordinate system is chosen in which one particle is at rest. The other particle is moving in a direction which, in the absence of the electrostatic force, would lead to its having a distance of closest approach d to particle 1. The actual track of particle 2 is a hyperbola and curves a, b and c show a sequence of tracks for increasing initial velocity of particle 2. The higher this velocity the closer is the distance of closest approach to d. If d is less than the range of nuclear forces and the velocity of particle 2 is high enough, there is then a possibility that a nuclear reaction will occur.

This classical description of the occurrence of nuclear reactions has been modified in one very important respect by the quantum theory. This states that there is a possibility that nuclear reactions will occur even when classical theory predicts that particles do not approach close enough for an interaction. This arises through Heisenberg's principle of uncertainty which states that it is impossible to give a precise value to both the position and momentum of a particle. There is an uncertainty δx in the position x and an uncertainty δp in the momentum p and these are related by:

$$\delta x \, \delta p \geqslant h/4\pi, \tag{4.6}$$

where h is Planck's constant. In a similar way Heisenberg's principle states that the energy of a particle cannot be measured in such a way that it and the time of measurement are known precisely. Thus there are uncertainties δE and δt in energy and time such that

$$\delta E \, \delta t \geqslant h/4\pi. \tag{4.7}$$

If the extra energy above the classical value that an incoming particle requires to approach within the range of nuclear forces and the time for which it requires that energy satisfy (4.7), there is a chance of a nuclear reaction occurring.

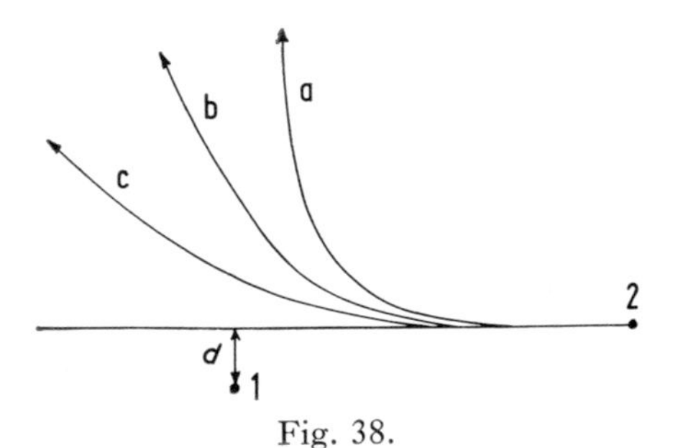

Fig. 38.

As a result the rate at which nuclear reactions occur is greater than would be expected on a classical theory. Before the quantum theory of nuclear reactions had been developed, it was difficult to see how nuclear reactions could produce the amount of energy which stars were radiating. At the same time it was difficult to envisage a satisfactory alternative source of energy, although the possibility of complete annihilation of matter into energy was considered. It was at this time that Eddington made his famous remark that, if the centre of the sun was not hot enough for the nuclear physicists, they must find a hotter place.

Inside a star the velocities possessed by particles are those of their random thermal motions. According to the kinetic theory of gases, the mean speed of a particle, $\bar{v}$, is given by the relation:

$$1{\cdot}086\bar{v} = (3kT/m)^{1/2} \text{ m s}^{-1}, \tag{4.8}$$

where as before m is the mean mass of the particles in the gas. This means that, if any of the particles are to have high enough velocities for nuclear reactions to occur, the temperature of the stellar gas must be high. The reactions are then known as *thermonuclear reactions.* We have already seen in the last chapter that, while the stellar gas remains perfect, its temperature will rise as the star loses energy (see the discussion on page 61). Thus it may be expected that, if the interior of a star is initially not at a high enough temperature for nuclear reactions to occur releasing energy, it will eventually become hot enough. At that stage what we have called hidden energy supplies on page 61 become available and the star can settle into a quasi-steady state in which the energy supplied by thermonuclear reactions provides the surface energy loss.

The higher the electric charges of interacting nuclei, the greater is the repulsive force between them and the higher the temperature of the stellar gas must be before thermonuclear reactions occur. The highly charged nuclei are also the more massive nuclei and this means that nuclear reactions between light elements occur at lower temperatures than nuclear reactions between heavy elements. Thus in an individual star it may be expected that the light elements will gradually be converted into heavier elements as the star evolves and its internal temperature rises, until finally the material has been converted into elements in the neighbourhood of iron in the periodic table. Once this has happened no more energy can be released by nuclear fusion reactions. It is possible that fusion reactions will not go this far. The proof in Chapter 3 that the stellar central temperature will rise as the star loses energy depends on the material of the star remaining a perfect gas. We shall see later in this chapter that, if the stellar gas becomes imperfect, it may be possible for the star's temperature to pass through a maximum and for the star to cool down and *die*. The final stages in the evolution of such stars will be discussed in Chapter 8.

As well as depending on the temperature of the stellar material, the rate of nuclear reactions clearly also depends on the density, but in this case the dependence is very simple. Thus, for the simplest two particle nuclear reactions in which two nuclei combine to form a third nucleus, probably with the emission of a photon, the energy release per unit volume is proportional to the product of the numbers of the two interacting particles in unit volume. That is, if nucleus A combines with nucleus B to form nucleus C and a photon γ through the reaction

$$\mathrm{A}+\mathrm{B}\rightarrow\mathrm{C}+\gamma\dagger, \tag{4.9}$$

the number of reactions occurring is proportional to $n(\mathrm{A})n(\mathrm{B})$, where $n(\mathrm{A})$, $n(\mathrm{B})$ are the numbers of particles of types A and B in unit volume.

† In what follows we shall sometimes use an abbreviated notation for such reactions, A(B, γ)C, where the initially present particles are written to the left of the comma and the final particles to the right. The notation γ is used for a photon as in these nuclear reactions the photons are γ-rays.

If the chemical composition is fixed, this means that the rate of energy generation per unit volume is proportional to ρ^2 and the rate of energy generation per unit mass, ε, is thus proportional to ρ. These two particle nuclear reactions are usually more important in stellar interiors than reactions involving three or more particles. This is true because the probability of more than two particles being simultaneously close enough together for a reaction involving all of them to take place is very small indeed, unless the density of the material is extremely high. It will, however, soon be seen that there is one very important three particle reaction which leads to an energy release proportional to ρ^2.

Nuclear reaction rates

From what we have said earlier it should be clear that the probability of a nuclear reaction occurring can be written as a product of two factors. These are the probability of two particles approaching close enough for the nuclear force to be important and the probability that a nuclear reaction will then occur. The first factor depends only on the masses and charges of the two particles, the number of particles present and the temperature. This factor is easy to calculate in principle, but requires a knowledge of quantum mechanics beyond the scope of this book. The second factor depends on the detailed properties of the two nuclei involved. It is not usually possible to calculate this factor and it must be determined from laboratory experiments.

Although we cannot calculate the rate at which thermonuclear reactions occur, it is perhaps useful to write down the formula to show how the reaction rate depends on T and the masses and charges of the particles. Suppose two interacting particles have masses $A_i m_H$ and $A_j m_H$, where m_H is the mass of the hydrogen atom and charges $q_i e$ and $q_j e$, where e is the electron charge. Suppose also that a fraction X_i by mass of the material is in the form of nucleus i and a fraction X_j in the form of nucleus j. Define the two quantities:

$$A = A_i A_j/(A_i + A_j)$$

and

$$\tau = 4{\cdot}25 \times 10^3 \, (q_i{}^2 q_j{}^2 A/T)^{1/3}. \qquad (4.11)$$

Then the number of reactions per kg per s involving the nuclei i and j can be written

$$R_{ij} \equiv C\rho \frac{X_i}{A_i}\frac{X_j}{A_j} \tau^2 \exp(-\tau)(A q_i q_j)^{-1}, \qquad (4.12)$$

where C is a constant depending on the particular properties of the nuclei concerned.

In the expression (4.12) the dependence of the reaction rate on density, temperature, nuclear abundances and nuclear masses and charges is shown clearly. When T is small, τ is large and the term $\exp(-\tau)$

leads to a very small reaction rate. As τ decreases, the reaction rate increases rapidly through the exponential term but this increase does not continue for ever. Eventually the term in τ^2 becomes more important than the exponential term when the temperature is very high and the reaction rate drops again. In practice we find that we are interested in temperatures at which there is still a rising trend in the reaction rate. The decrease in reaction rate as the charges on the interacting nuclei are increased is also apparent from expressions (4.11) and (4.12) as there is a strong dependence on q_i and q_j in the exponential term.

Hydrogen burning reactions
It is now believed that the most important series of reactions occurring in stars are those converting hydrogen into helium. This is known as *hydrogen burning*. An important feature of these reactions is that they involve the conversion of two protons into neutrons for each nucleus of ^4He (α particle) produced. As was mentioned on page 83, the conversion of a proton into a neutron requires the operation of the weak nuclear interaction. Thus hydrogen burning involves rather more than two particle nuclear reactions of the type which has just been described. Two basic reaction chains have been proposed: the proton–proton (PP) chain and the carbon–nitrogen (CN) cycle.

In the first chain hydrogen is converted directly into helium, but in the CN cycle nuclei of carbon and nitrogen are used as catalysts.

Details of the PP chain are shown in expressions (4.13)–(4.15) and of the CN cycle in expression (4.16). The notation used has been explained in the footnote on page 86. The proton–proton chain divides into three main branches which are called the PP I, PP II and PP III chains. The first reaction is the interaction of two protons to form a

PP I Chain

$$\left.\begin{array}{ll} 1 & p(p,\, e^+ + \nu)d \\ 2 & d(p,\, \gamma)^3He \\ 3 & ^3He(^3He, p + p)^4He \end{array}\right\} \quad (4.13)$$

nucleus of heavy hydrogen (deuteron, d) with the emission of a positron (e^+) and a neutrino (ν). The deuteron then captures another proton and forms the light isotope of helium with the emission of a γ-ray. At this stage two important possibilities arise. The nucleus of ^3He can either interact with another nucleus of ^3He or with an α particle, which has already been formed, or may have been present initially if the star contained helium at birth.

In the former case we have the final reaction of the PP I chain while the latter reaction leads into either the PP II or the PP III chain. The remainder of the reactions will not be described in detail, but it should be noted that there is another choice in the chain when ^{7}Be either captures an electron to form ^{7}Li or captures another proton to form ^{8}B. At the end of the PP III chain, the unstable nucleus ^{8}Be breaks up to form two α particles.

PP II Chain

This starts with reactions 1 and 2.

$$\left.\begin{array}{lll} \text{Then} & 3' & {}^{3}He({}^{4}He, \gamma){}^{7}Be \\ & 4' & {}^{7}Be(e^{-}, \nu){}^{7}Li \\ & 5' & {}^{7}Li(p, \alpha){}^{4}He \end{array}\right\} \quad (4.14)$$

PP III Chain

This starts with reactions 1, 2 and 3′.

$$\left.\begin{array}{lll} \text{Then} & 4'' & {}^{7}Be(p, \gamma){}^{8}B \\ & 5'' & {}^{8}B(, e^{+} + \nu){}^{8}Be \\ \text{and} & 6'' & {}^{8}Be \rightarrow 2{}^{4}He \end{array}\right\} \quad (4.15)$$

CN Cycle

$$\left.\begin{array}{ll} 1 & {}^{12}C(p, \gamma){}^{13}N \\ 2 & {}^{13}N(, e^{+} + \nu){}^{13}C \\ 3 & {}^{13}C(p, \gamma){}^{14}N \\ 4 & {}^{14}N(p, \gamma){}^{15}O \\ 5 & {}^{15}O(, e^{+} + \nu){}^{15}N \\ 6 & {}^{15}N(p, {}^{4}He){}^{12}C \end{array}\right\} \quad (4.16)$$

The important feature of the CN cycle is that it starts with a carbon nucleus to which are added four protons successively. In two cases the proton addition is followed immediately by a β-decay, with the emission of a positron and a neutrino, and at the end of the cycle a helium nucleus is emitted and a nucleus of carbon remains. The carbon and nitrogen act purely as catalysts in these reactions and are neither produced nor destroyed. Actually in both the PP chain and the CN cycle there are some less important side reactions which have not been listed. If, of course, any stars exist which do not contain any carbon or nitrogen, the CN cycle cannot occur and all hydrogen burning must be through the PP chain. However, a very small amount of carbon will suffice to make the CN cycle important in some stars as we shall see when we discuss how the energy release from the two reaction chains depends on ρ, T and chemical composition.

Neutrinos

These hydrogen burning reactions are of particular interest because the conversion of a proton into a neutron involves the emission of not only a positron but also a neutrino. The neutrino is a particle which is apparently without mass, possesses no electromagnetic properties (charge, magnetic moment etc.) and does not take part in strong interactions. In fact it was originally suggested that it must exist because otherwise energy and momentum would not be conserved in β-decay reactions. Rather than jettison apparently well-established conservation laws, it was felt preferable to hypothesize that an additional, as yet undetected, particle was emitted in the reaction.

After more than 20 years the neutrino (or more accurately the anti-neutrino) was detected in 1953. Neutrinos interact very weakly with other matter and a neutrino of 1 MeV energy would pass through about 10 parsecs of water without being seriously deflected or absorbed. It is thus very difficult to detect an individual neutrino but, since there is a small, but finite, probability that a neutrino will be absorbed in a short distance, neutrinos can be detected provided a sufficiently strong flux of neutrinos can be obtained. In the first experiments the anti-neutrinos were emitted by β-decay of unstable nuclei; inside the nucleus an individual neutron decayed through reaction (4.5):

$$\mathrm{n} \rightarrow \mathrm{p} + \mathrm{e}^- + \bar{\nu}.$$

The anti-neutrinos were then captured by protons through a reaction which is essentially the inverse of (4.5):

$$\bar{\nu} + \mathrm{p} \rightarrow \mathrm{n} + \mathrm{e}^+. \tag{4.17}$$

Subsequently the neutron decayed through reaction (4.5) and the positron annihilated with an electron to produce γ-rays:

$$\mathrm{e}^+ + \mathrm{e}^- \rightarrow \gamma + \gamma. \tag{4.18}$$

The neutron decay and the gamma ray production were both observed and the near coincidence of these events enabled occurrence of the anti-neutrino capture reaction to be deduced. It is hardly surprising that the positive detection of the anti-neutrino was not obtained easily.

Almost all of the neutrinos emitted by nuclear reactions in the centre of a star escape from the star without any further interaction, whereas, as will be seen later in this chapter, the low energy γ-rays emitted in the nuclear reactions only travel a small fraction of the star's radius before they are absorbed. The neutrinos in reactions (4.13)–(4.16) carry away between 2% and 6% of the energy released in the reactions and this energy is lost almost immediately and is of no use to the star. Thus the period of hydrogen burning is a few per cent shorter than it would otherwise have been. Perhaps the most interesting property of these neutrinos is that they are potentially capable of giving the observer on Earth some information about conditions in the centre of the Sun whereas photons only give direct information about the surface layers. Although neutrinos interact only very weakly with matter there is a possibility that a few can be detected and can give some direct information about the region in which nuclear reactions are occurring in the Sun.

For some time an experiment has been under way to try to detect these neutrinos in a large tank containing 400 000 litres of cleaning fluid (perchloroethylene C_2Cl_4) placed almost a mile underground in a gold mine! The neutrinos are absorbed by ^{37}Cl, the heavy isotope of chlorine, through the reaction

$$^{37}\mathrm{Cl} + \nu \rightarrow {}^{37}\mathrm{Ar} + \mathrm{e}^{-}. \tag{4.19}$$

The radioactive argon is then separated by a chemical process from the remainder of the fluid and the number of radioactive atoms produced is counted by observing the reverse reaction:

$$^{37}\mathrm{Ar} \rightarrow {}^{37}\mathrm{Cl} + \mathrm{e}^{+} + \nu. \tag{4.20}$$

The experiment is placed deep in a mine to minimize the production of ^{37}Ar by other agents such as cosmic rays. At the time of writing, the detection rate of neutrinos appears to be rather less than theory predicts, but it is not at present clear that there is a significant discrepancy. Further comments on this experiment are made in Chapter 6.

Energy release from hydrogen burning

Using experimental or, in many cases, extrapolated values for the rates at which the reactions in the chains (4.13)–(4.16) occur, it is possible to tabulate the energy release of the CN cycle and the total release from the three branches of the PP chain as a function of temperature and this is shown in fig. 39. In both cases the energy release per kg is proportional to the density. The energy release from the PP chain is proportional to the square of the fractional hydrogen content X_H while that for the CN cycle is proportional to the product of the hydrogen concentration and the carbon concentration X_C. The diagram is drawn for a value of X_C/X_H appropriate to population I stars like the Sun which are relatively rich in elements heavier than hydrogen and helium. A change in the value of X_C/X_H only leads to a bodily displacement of one curve relative to the other.

Consider the implication of this diagram for the Sun. The Sun's energy release is at the rate of 2×10^{-4} W kg^{-1} averaged over its whole mass. The mean density of the Sun is $1{\cdot}4 \times 10^{3}$ kg m^{-3} and the observations of the chemical composition of the solar surface suggest that $X_H \simeq \frac{3}{4}$. If the entire Sun were at the same temperature and density, it would have a value of $\varepsilon/\rho X_H^2$ of about $2{\cdot}5 \times 10^{-7}$ and it can be seen from the figure that the temperature would be about 5×10^6 K. In fact the flow of heat outwards from the centre of the Sun requires the central temperature to be higher than the mean temperature so that this estimate is probably somewhat low. This estimate of the interior temperature of the Sun can be compared with a lower limit to the mean temperature of 2×10^6 K which was obtained from expression (3.32). Although the estimate we have made here is very rough, it is satisfactory that hydrogen burning reactions will supply energy at the observed rate for stellar interior temperatures which are quite similar to those estimated in Chapter 3.

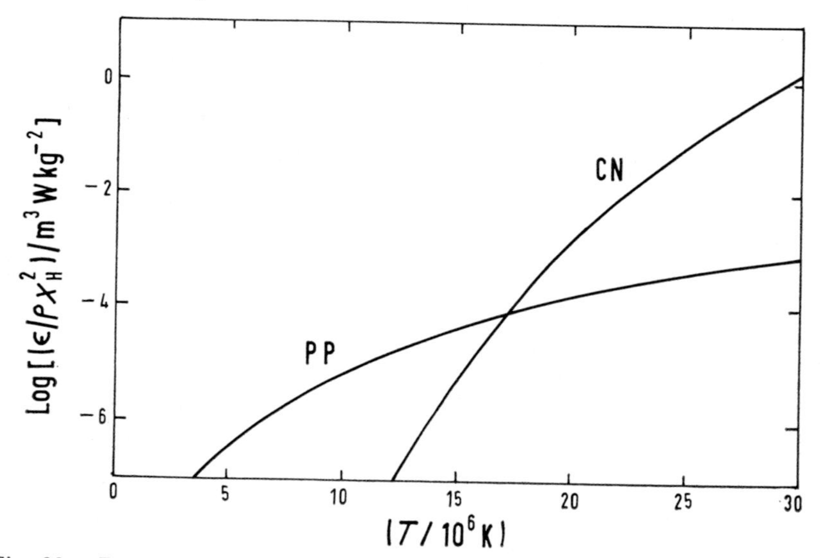

Fig. 39. Rate of energy release from hydrogen burning as a function of temperature for the two reaction chains.

The energy releases by the PP chain and CN cycle are smooth functions of temperature. In a limited temperature range, we can replace the true dependence on temperature shown in fig. 39 or equation (4.11) by a power law obtained by replacing the curve by its tangent, or perhaps better by an appropriate chord parallel to the tangent at some point in the range. Thus, in a region not far below the temperature where the PP chain and CN cycle are equally important, we can represent the energy generation rate by:

$$\varepsilon_{\rm PP} \simeq \varepsilon_1 X_{\rm H}^2 \rho T^4, \tag{4.21}$$

while for the CN cycle at a slightly higher temperature:

$$\varepsilon_{\rm CN} \simeq \varepsilon_2 X_{\rm H} X_{\rm C} \rho T^{17}, \tag{4.22}$$

where in (4.20) and (4.21) ε_1 and ε_2 are constants.

Clearly the true law of energy release is not a power law, but this is quite a good approximation as the energy release increases very rapidly with temperature and the range of temperature in which significant release occurs is small. In the next chapter we shall see that the use of approximate laws of energy production of the form

$$\varepsilon = \varepsilon_0 \rho T^\eta \tag{4.23}$$

enables us to obtain useful qualitative information about the structure of stars.

Other nuclear reactions involving light elements

It has been stated several times earlier that stars composed of a perfect gas heat up as they radiate and that this heating up proceeds until thermonuclear reactions start in the interior. It has also been stated that reactions involving the nuclei of lowest charge occur first so that it might be expected that no significant nuclear reactions can occur at a lower temperature than that at which hydrogen burns to helium. This is not quite true because both the PP chain and CN cycle are atypical. In each case weak interactions must occur to change protons into neutrons and these weak interactions are always slower than strong interactions. In addition, each chain has another peculiarity. The first reaction of the PP chain really involves two steps:

$$\mathrm{p}+\mathrm{p}\rightarrow {}^2\mathrm{He}\rightarrow \mathrm{d}+\mathrm{e}^+ +\nu. \tag{4.24}$$

The intermediate nucleus ^{2}He is highly unstable and it usually decays back to two protons rather than β-decaying into a deuteron. This means that reaction (4.24) occurs very infrequently. In the case of the CN cycle the carbon and nitrogen nuclei have relatively high electric charges and the reaction rate is lower than it would be if all the nuclei involved had low electric charge.

In fact deuterium, lithium, beryllium and boron will burn at lower temperatures than hydrogen because they can all burn without β-decays and without involving any particles with charges as high as carbon, nitrogen and oxygen. Bearing in mind that hydrogen is almost always the most abundant element, deuterium and the stable isotopes of lithium, beryllium and boron are destroyed by reactions involving protons. The actual reactions are:

$$\left.\begin{array}{l} \mathrm{d}(\mathrm{p},\ \gamma)^3\mathrm{He}, \\ ^6\mathrm{Li}(\mathrm{p},\ ^3\mathrm{He})^4\mathrm{He}, \\ ^7\mathrm{Li}(\mathrm{p},\ \gamma)^8\mathrm{Be}\rightarrow 2\,^4\mathrm{He}, \\ ^9\mathrm{Be}(\mathrm{p},\ ^4\mathrm{He})^6\mathrm{Li}(\mathrm{p},\ ^3\mathrm{He})^4\mathrm{He}, \\ ^{10}\mathrm{B}(\mathrm{p},\ ^4\mathrm{He})^7\mathrm{Be}, \\ ^{11}\mathrm{B}(\mathrm{p},\ \gamma)3\,^4\mathrm{He}, \end{array}\right\} \tag{4.25}$$

and the ^{7}Be produced in the fifth reaction is destroyed as in the PP chain. Although these reactions are expected to be the first that occur in stars and although they individually have a high yield of energy, they are not expected to play an important role in stellar structure (except that the first and third occur in the PP chain when the deuterium and lithium have themselves been built in the chain) because it is not believed that the elements concerned ever have high abundances. Thus we shall continue to regard the hydrogen burning reactions as the first significant ones to occur after a star is born.

It should be remarked that it is reactions involving deuterium rather than protons which are used in the hydrogen bomb and in experiments attempting to control thermonuclear reactions in the laboratory. Thus the suggested reaction chains for controlled thermonuclear reactions are:

$$\begin{aligned} &\mathrm{d(d,\,p)^3H(d,\,n)^4He} \\ &\mathrm{(d,\,n)^3He(d,\,p)^4He.} \end{aligned} \qquad (4.26)$$

Recent estimates of the possibility of obtaining a useful energy release from controlled thermonuclear reactions suggest that it may be necessary for the initial fuel to be a mixture of deuterium and tritium (^3H), so that only one reaction in the above set is used.

Helium burning reactions

If it is accepted that the burning of hydrogen will be the first nuclear process of importance in stellar evolution, there will come a time when there is no longer any hydrogen left in the central regions of the star. At this stage the central regions will still be hotter than the outer regions and thus energy will continue to flow outwards. As there is no longer any nuclear energy release, the energy flow can only come from the thermal energy. Any loss of thermal energy reduces the pressure of the stellar gas and the central regions are then compressed by the overlying layers. Provided the material remains a perfect gas, this compression leads to a rise in temperature and the rise in temperature continues until the next significant nuclear reactions occur; this may also be true even if the centre of the star does not remain a perfect gas.

The next significant reactions to occur are those involving ^4He, the product of hydrogen burning. It might be expected that the important reaction would involve the combination of two helium nuclei to produce either one or two other nuclei. Unfortunately this does not work as ^8Be is very unstable and decays to two helium nuclei again, as we have already seen in the reactions of the PP chain, while the other reactions involving two helium nuclei require energy rather than releasing energy. For some time it was unclear how further nuclear burning would proceed. Then it was realized that very rarely a third helium nucleus could be added to ^8Be before it decayed. Carbon is then formed by the chain:

$$\begin{aligned} &{}^4\mathrm{He} + {}^4\mathrm{He} \rightarrow {}^8\mathrm{Be}, \\ &{}^8\mathrm{Be} + {}^4\mathrm{He} \rightarrow {}^{12}\mathrm{C} + \gamma. \end{aligned} \qquad (4.27)$$

As with hydrogen burning reactions, this is an unusual reaction.

Helium burning is effectively a three particle reaction so that the energy release per kg is proportional to the square of the density instead of being linearly proportional to density as in the case of hydrogen burning. The reaction rate is again very strongly dependent on temperature. In stars helium burning occurs typically at temperatures of about 10^8 K and near this temperature the rate of energy release is:

$$\varepsilon_{3\mathrm{He}} \simeq \varepsilon_3 X_{\mathrm{He}}^3 \rho^2 T^{40}, \tag{4.28}$$

where ε_3 is a constant and X_{He} is the fractional concentration of helium by mass.

Once helium is used up in the central regions of a star, further contraction and heating may occur and that may lead to additional nuclear reactions such as the burning of carbon. For the present we will not discuss these reactions, but will remark again that the majority of the possible energy release by nuclear fusion reactions has occurred by the time that hydrogen and helium have been burnt.

Opacity

We now turn to a discussion of the opacity of stellar material. In Chapter 3 it has been stated that the flow of energy by conduction and radiation is essentially similar in nature and that, inside a star, the rate at which energy flows by these processes is determined by one quantity, the opacity κ. Thus we have equation (3.51):

$$\frac{\mathrm{d}T}{\mathrm{d}r} = -\frac{3\kappa L\rho}{16\pi acr^2T^3},$$

which relates the rate of energy transport to the temperature gradient and the opacity. In the last chapter no formula was given for κ and a full discussion of how the formula is obtained is outside the scope of the present book. We will, however, discuss the principles underlying the determination of κ.

It has already been mentioned in Chapter 2 that, if matter and radiation are in equilibrium with one another at a temperature T, the radiation present is entirely described in terms of the Planck function $B_\nu(T)$ where

$$B_\nu(T) = \frac{2h\nu^3}{c^2}\,\frac{1}{\exp\,(h\nu/kT)-1}. \tag{2.5}$$

$B_\nu(T)$ is the energy crossing an area of a square metre per second in a unit frequency range and unit solid angle. In thermal equilibrium there is an equal flow or radiation in all directions. Inside a star conditions cannot be precisely those of thermal equilibrium because in that case there would be no net flow of energy in the radial direction. However, the departures of the radiation intensity from the Planck function are very small indeed inside a star.

The opacity of stellar material is a measure of the resistance of the material to the passage of radiation; equivalently the radiative conductivity measures the ease with which energy flows. The probability that an individual photon will be absorbed depends on its frequency. Thus we can define the monochromatic mass absorption coefficient, κ_ν, of the material, which is such that radiation of intensity I_ν is changed by δI_ν in a distance δx, where

$$\delta I_\nu = -\kappa_\nu \rho I_\nu \delta x. \tag{4.29}$$

The dimensions of κ_ν are $\mathrm{m^2\ kg^{-1}}$ and it is called a mass absorption coefficient because of the inclusion of ρ in the definition (4.29). When considering the radiative conductivity of stellar material, it is reasonable to suppose that the effective conductivity will depend mainly on the conductivity in a frequency range in which the number of photons is a maximum. This means that, in forming the average conductivity, the conductivity at any frequency should be multiplied by a quantity depending on the number of photons of that frequency which are present, before the average is calculated. As the opacity is essentially a reciprocal of the conductivity, this means that the reciprocal of the absorption coefficient, κ_ν, should be weighted with the number of photons present in forming the reciprocal of the opacity.

As mentioned above, energy only flows because the temperature is higher near the centre of the star. At any point the outward flowing radiation has been emitted at a slightly higher temperature and has a frequency distribution approximating a Planck function at that higher temperature. Similarly the inward flowing radiation has a frequency distribution corresponding to a Planck function at a slightly lower temperature. The net radiative conductivity is then obtained by multiplying the conductivity at any frequency by the difference between these two Planck functions, by integrating over frequency and by suitably normalizing the answer. In terms of opacity rather than conductivity

$$\frac{1}{\kappa} = \int_0^\infty \frac{1}{\kappa_\nu} \frac{\mathrm{d}B_\nu}{\mathrm{d}T} \mathrm{d}\nu \Big/ \int_0^\infty \frac{\mathrm{d}B_\nu}{\mathrm{d}T} \mathrm{d}\nu. \tag{4.30}$$

The effective opacity is calculated from this formula once the absorption coefficient at all frequencies is known.

The sources of opacity

The discussion of stellar opacity now involves considering all of the microscopic processes which contribute to the absorption of radiation of frequency ν. It is impossible to discuss these in detail in this book, but an account can be given of the basic types of process involved. There are four of these:

(i) bound–bound absorption,
(ii) bound–free absorption,
(iii) free–free absorption,
(iv) scattering.

These processes are described below. The first three are called true absorption processes because they involve the disappearance of a photon, while in the fourth case only the direction of motion of the photon is altered. In addition there is the contribution to the effective opacity from thermal conduction. This will not be discussed explicitly below, but it is included in the numerical calculations which are mentioned.

Bound–bound absorption
In this case an electron is moved from one bound orbit in an atom or ion to an orbit of higher energy with the absorption of a photon (fig. 40).

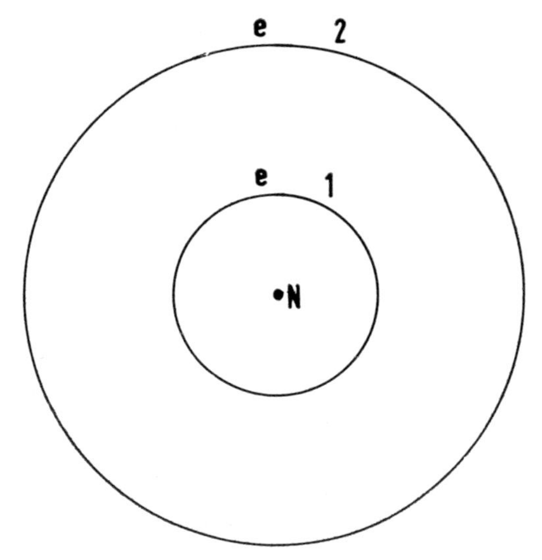

Fig. 40. Bound–bound absorption. An electron e can move from orbit 1 to orbit 2 about a nucleus N, with absorption of a photon.

Thus, if the energy of the electron in the two orbits is E_1 and E_2 respectively, a photon of frequency ν_{BB} can produce the transition if

$$E_2 - E_1 = h\nu_{BB}. \tag{4.31}$$

These bound–bound processes, which are responsible for the characteristic spectral lines of different elements and which lead to the occurrence of spectral lines in the visible radiation from stars, are not very important in the deep interiors for two reasons. As all of the atoms are highly ionized, only a small minority of electrons are in bound states. In addition the majority of photons have energy somewhere near that corresponding to the maximum of the Planck function (2.5), which occurs at frequency ν_{max} satisfying:

$$h\nu_{max}/kT = 2{\cdot}82. \tag{4.32}$$

For conditions in deep stellar interiors the energy $h\nu_{max}$ ($= 2{\cdot}82kT$) is greater than the separation in energy between atomic bound states and the photons are more likely to cause a bound–free absorption which we now describe.

Bound–free absorption

This involves an electron in a bound state around a nucleus being moved into a free hyperbolic orbit by the absorption of a photon (fig.

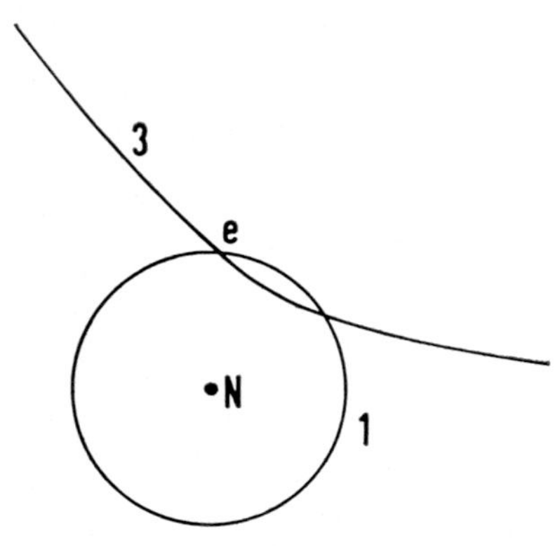

Fig. 41. Bound–free absorption. The absorption of a photon can move an electron from the bound orbit 1 to the free orbit 3.

41). A photon of frequency ν_{BF} can be absorbed and convert a bound electron of energy E_1 into a free electron of energy E_3 provided that

$$E_3 - E_1 = h\nu_{BF}. \quad (4.33)$$

In this case, as in the case of bound–bound absorption, the importance of the process is reduced because of the scarcity of bound electrons. However, provided the photon has sufficient energy to remove the electron from the atom, any value of energy can lead to a bound–free process.

Free–free absorption

In this case the electron is initially in a free state with energy E_3 and it absorbs a photon of frequency ν_{FF} and moves to a state with energy E_4, where

$$E_4 - E_3 = h\nu_{FF}. \quad (4.34)$$

There is no restriction on the energy of a photon which can induce a free–free transition but, in both free–free and bound–free absorption, it is found that low energy photons are more likely to be absorbed than high energy photons.

Scattering
Finally it is possible for a photon to be scattered by an electron or an atom. On a classical picture, this process can be idealized as a collision between two particles which bounce off one another. If the energy of the photon satisfies

$$h\nu \ll mc^2, \tag{4.35}$$

where m is the mass of the particle doing the scattering, the particle is scarcely moved by the collision. In this case the photon can be imagined to be bounced off a stationary particle (fig. 42). The inequality is valid throughout most stars, but is violated in stars with extremely

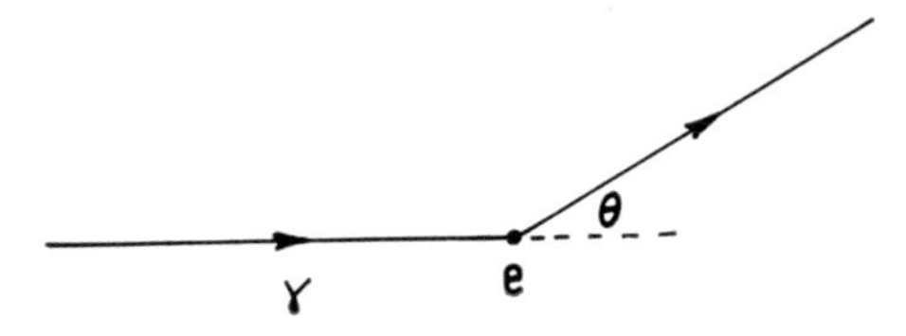

Fig. 42. The scattering of a photon γ by an electron e.

high internal temperatures. Although this process does not lead to the true absorption of radiation it does slow down the rate at which energy escapes from a star because it continually changes the direction of the photons.

The calculation of stellar opacity is a very complicated process as all atoms and ions must be considered. Because expression (4.30) for κ involves what is known as a *harmonic mean* of κ_ν rather than a direct mean, we cannot calculate a mean absorption coefficient for each chemical element independently and then add the results together to form κ. Instead all of the contributions to κ_ν must be added together before the mean is calculated. This means that each time we wish to consider a star of a different chemical composition, expression (4.30) must be calculated afresh. Another property of the expression (4.30) for κ is that a sensible value of κ can only be obtained if we have an estimate of the monochromatic absorption coefficient at all frequencies. If, in expression (4.30), κ_ν is put equal to zero in any frequency band, the resulting value of κ is zero, which suggests that radiation escapes freely. This is clearly not true, if κ_ν is non-zero for most frequencies.

Numerical values for opacity
Further details of the calculation of opacities cannot be given here, but the results of a recent calculation are shown in fig. 43. For material of one particular chemical composition, the opacity is shown as a function of temperature and density. It can be seen that the opacity is low at both very high and low temperatures. At high temperatures most of the photons have high energy and, as has been mentioned

previously, they are absorbed less easily than lower energy photons. At low temperatures most atoms are not ionized and there are few electrons available to scatter radiation or to take part in free–free absorption processes, while most photons have insufficient energy to ionize atoms. The opacity has a maximum at intermediate temperatures where bound–free and free–free absorption are very important.

From fig. 43, one result of immediate interest can be obtained. In the Sun the central density is about 10^5 kg m^{-3} and the opacity is

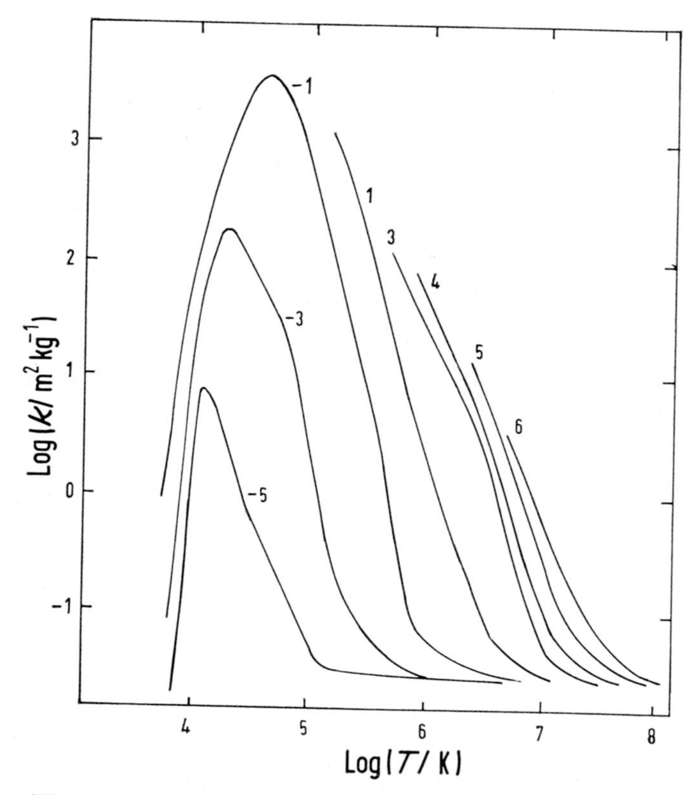

Fig. 43. The opacity κ as a function of temperature and density. Each curve represents a different value of the density and is labelled by log (ρ/kg m^{-3}).

about 10^{-1} m^2 kg^{-1}, so that $\kappa\rho \simeq 10^4$ m^{-1}. From equation (4.29), this means that a typical photon travelling from the centre of the Sun is absorbed or scattered when it has travelled about 10^{-4} m. Further out in the Sun, when the density is about 10^3 kg m^{-3}, the opacity is 10 m^2 kg^{-1}, so that once again the mean free path of radiation is about 10^{-4} m. The central temperature of the Sun is believed to be about $1{\cdot}5 \times 10^7$ K so that the mean temperature gradient in the Sun is about

2×10^{-2} K m^{-1}. This means that a typical temperature difference between the point at which a photon is emitted and the point at which it is absorbed is 2×10^{-6} K. This is remarkably small and this is the reason why the intensity function I_ν is able to remain so very close to the Planck function $B_\nu(T)$ in stellar interiors. The departure from true thermal equilibrium is very slight.

Approximate form for opacity

Approximate analytical expressions for the opacity in particular ranges of temperature and density can be read off from the curves of fig. 43. At high temperatures, there is a range of density for which the opacity is scarcely dependent on either temperature or density (although there is a dependence on chemical composition which cannot be shown in fig. 43), so that to a first approximation we may write:

$$\kappa = \kappa_1, \tag{4.36}$$

where κ_1 is a constant for stars of a given chemical composition. At high temperatures the main source of opacity is the scattering of radiation by free electrons (Compton scattering) and, if no other processes are important, the opacity has precisely the form (4.36).

At lower temperatures the processes of bound–free and free–free absorption become important and there is a range of temperature in which the opacity increases with increasing density and with decreasing temperature. A reasonable analytical approximation to the opacity has the form:

$$\kappa = \kappa_2 \rho / T^{3.5}, \tag{4.37}$$

where κ_2 is again a constant for stars of given chemical composition. At even lower values of the temperature, the opacity decreases with decreasing temperature and an approximate analytical form for the opacity in that regime is:

$$\kappa = \kappa_3 \rho^{1/2} T^4, \tag{4.38}$$

where κ_3 is another constant. This region of the opacity curve is not important in stars whose surface temperature is high enough for the abundant elements, hydrogen and helium, to be significantly ionized even at the stellar surface. Figure 44 shows how these approximate expressions (4.36)–(4.38) fit to one of the calculated curves of fig. 43.

When calculations were first made of the structure and evolution of the stars, simple power law approximations to the law of energy release, such as:

$$\varepsilon = \varepsilon_0 \rho T^\eta, \tag{4.23}$$

and to the law of opacity:

$$\kappa = \kappa_0 \rho^{\lambda - 1} / T^{\nu - 3} \tag{4.39}$$

(where the choice of the exponents $\lambda - 1$, $\nu - 3$ is made to give a simple form in equation (3.51), and v must not be confused with the frequency) were extremely important. Their use enabled progress to be made in the subject without the use of electronic computers, which had not yet been developed. Indeed, as we shall see in the next chapter, quite useful results could be obtained without the aid of any type of calculating

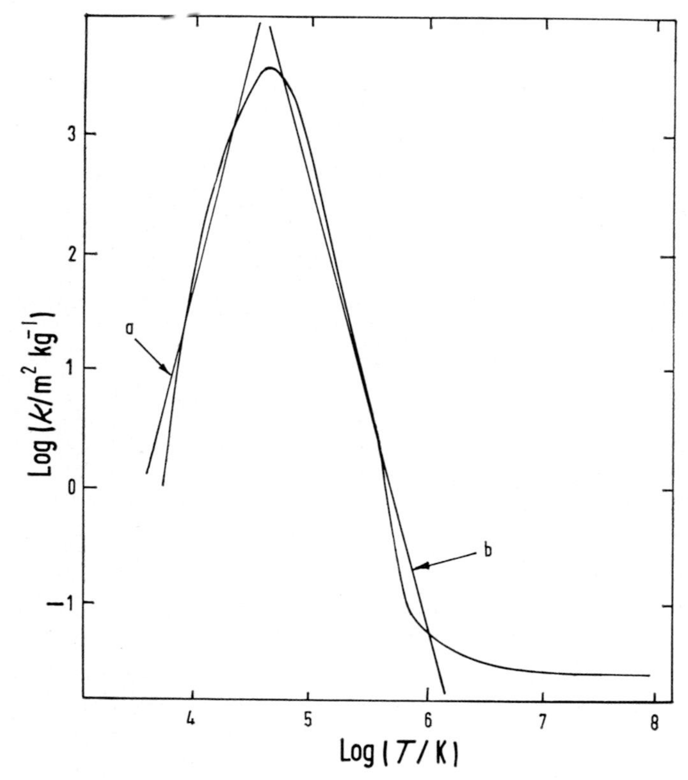

Fig. 44. The fit of approximate expressions for the opacity to the true expression. The curve is for a density of 10^{-1} kg m^{-3}. The straight line a is the approximation (4.38), the straight line b is the approximation (4.37) and the opacity approaches a constant value, approximation (4.36), at high temperature.

machine. Today the development of large electronic computers enables much more detailed information about the laws of opacity and energy generation to be used in the computations.

The equation of state of stellar material

The third quantity whose behaviour must be considered is the pressure. We have already stated that stellar material is gaseous and that in many cases it behaves like a perfect gas. If it is a perfect gas, the gas pressure is:

$$P_{gas} = nkT, \tag{3.25}$$

where n is the number of particles per cubic metre and k is Boltzmann's constant. To obtain this in the form

$$P_{gas} = P(\rho, T, \text{composition}),$$

we require an expression for n in terms of ρ, T and composition. Astronomers usually write equation (3.25) in an alternative form by introducing:

$$\mu \equiv \rho/nm_H, \tag{4.40}$$

where m_H is the mass of the hydrogen atom ($1{\cdot}67 \times 10^{-27}$ kg), so that μ is the mean mass of the particles in the gas in terms of the mass of the hydrogen atom. μ is called the mean molecular weight of the stellar material. Then introducing the gas constant:

$$\mathscr{R} = k/m_H, \tag{4.41}$$

where $\mathscr{R} = 8.3 \times 10^3$ J K^{-1} kg^{-1}†, equation (3.25) takes the form:

$$P_{gas} = \mathscr{R}\rho T/\mu, \tag{4.42}$$

which is the form we shall use subsequently in this book.

Mean molecular weight of ionized gas

We now require an expression for μ as a function of ρ, T and chemical composition. The calculation of μ for completely general values of ρ and T is very complicated because to obtain a value for n the fractional ionization of all the elements has to be computed. Fortunately, throughout most of the interior of stars, the expression can be simplified for two reasons. In the first place all of the elements are highly ionized and, secondly, hydrogen and helium are very much more abundant than all of the other elements and they are certainly fully ionized in stellar interiors. This means that only a very small error is made in the value of μ if it is assumed that all of the material is fully ionized. Near the stellar surface this approximation becomes inadequate and a more careful discussion of the value of μ is required. This is important for detailed calculations of stellar structure, but for our approximate discussion in Chapter 5 only the value of μ for a fully ionized gas will be required.

If the material is assumed to be fully ionized, the calculation of μ proceeds as follows. Suppose that the fractional abundances by mass of hydrogen, helium and all elements other than hydrogen and helium are called X, Y and Z (instead of X_H, X_{He}, etc. as defined above), so that

$$X + Y + Z = 1. \tag{4.43}$$

† Note that $\mathscr{R}$ differs from R by a factor of approximately 10^3. They are essentially equal in c.g.s. units, but differ in SI units because, while the unit of mass is different, the mass of a mole is not.

This means that in a cubic metre of material of density ρ there is a mass $X\rho$ of hydrogen, a mass $Y\rho$ of helium and a mass $Z\rho$ of heavier elements.

In a cubic metre there are therefore $X\rho/m_H$ hydrogen atoms. Each ionized hydrogen atom consists of two particles, a proton and an electron, so that the hydrogen provides $2X\rho/m_H$ particles per cubic metre. In a similar way, as the mass of a helium atom is $4m_H$, there are $Y\rho/4m_H$ helium atoms per cubic metre. Each ionized helium atom consists of three particles and the helium provides $3Y\rho/4m_H$ particles per cubic metre. In principle each heavy element should be considered separately. However, for these heavier elements, the number of electrons is always about half the atomic mass number in units of m_H. Thus to a reasonable approximation when they are fully ionized they supply about one particle for every $2m_H$; the number is somewhat larger for atoms of low atomic weight and lower for the heavier atoms. To a first approximation the number of particles provided by the heavy elements is $Z\rho/2m_H$.

The total number of particles per cubic metre can then be written:

$$n = (\rho/m_H)[2X+3Y/4+Z/2]. \tag{4.44}$$

Using equation (4.43), expression (4.44) can be rewritten:

$$n = (\rho/4m_H)[6X+Y+2]. \tag{4.45}$$

This can be combined with equation (4.40) to give:

$$\mu = 4/(6X+Y+2), \tag{4.46}$$

which is a good approximation for μ except in the cool outer regions of a star. In many cases the fractional abundance of the heavy elements is so small that Z can be neglected in equation (4.46) and Y replaced by $1-X$ to give:

$$\mu = 4/(3+5X), \tag{4.47}$$

and this expression for μ is used in the next chapter.

Departures from perfect gas law

Expressions (4.46) and (4.47) give a good approximation to the mean molecular weight of a fully ionized gas and, while the gas remains perfect, its pressure can then be found from equation (4.42). We have mentioned earlier that as a star evolves it tends to contract and heat up and that this process continues while the stellar gas remains perfect. Is there any reason why it should not remain perfect? We can expect to find departures from the perfect gas law at high enough densities when the particles in the gas are packed close together, as in the case of the well-known van der Waals' forces.

In fact, in the ionized gas in stellar interiors, the first deviation occurs because electrons have to obey Pauli's exclusion principle. In its most familiar form this principle states that no more than one electron can occupy any one bound energy state in an atom†. The principle plays an important role in the arrangement of electrons in atoms and in the explanation of the periodic table of the elements. The Pauli exclusion principle also places a restriction on the relative position and momentum of two free electrons which are not attached to atoms. The nearer together two electrons are the greater must be the difference between their momenta and these differences must essentially satisfy Heisenberg's uncertainty principle:

$$\delta x \, \delta p \gg h/4\pi. \tag{4.6}$$

This means that if particles are closely packed together they may be forced to have a higher momentum than is predicted by the kinetic theory of gases for a perfect gas. As a result of this a gas at a given temperature and density has a higher total internal energy and a higher pressure than is predicted by the perfect gas law. A gas in which the Pauli exclusion principle is important is called a *degenerate gas*. Because at a given temperature ions have a higher momentum than electrons, the ions are less likely to be in danger of violating the Pauli exclusion principle. In stars electrons may form a degenerate gas, but the ions can almost always be treated as a perfect gas. The derivation of the formula for the pressure of a degenerate gas cannot be given here as it is too complicated, but the result of the calculation is as follows.

At a high enough density, the momentum of a particle is essentially determined by the Pauli exclusion principle rather than by the temperature of the gas. This means that the pressure and internal energy of the gas become essentially independent of temperature. The precise form of the pressure depends on whether the highest momentum possessed by a particle is greater than, or less than, $m_e c$, where m_e is the electron mass and c the velocity of light. The maximum possible velocity of an electron is, of course, c but according to the special theory of relativity the momentum, p, can exceed $m_e c$ as it is given by the formula:

$$p = m_e v/(1 - v^2/c^2)^{1/2}, \tag{4.48}$$

which shows that p increases without limit as v approaches c.

If the maximum electron momentum, p_0, satisfies $p_0 \ll m_e c$, the pressure can be shown to be:

$$P_{\text{gas}} \simeq K_1 \rho^{5/3}, \tag{4.49}$$

where

$$K_1 = \frac{h^2}{2m_e}\left(\frac{3}{\pi}\right)^{2/3}\left(\frac{1+X}{2m_H}\right)^{5/3}. \tag{4.50}$$

As before, X is the mass fraction in the form of hydrogen and the gas has

† In fact if the property known as electron spin is allowed for, *two* electrons with oppositely directed spin can be in any one state.

been assumed to be fully ionized. If the momentum p_0 satisfies $p_0 \gg m_e c$, the pressure is given by:

$$P_{gas} \simeq K_2 \rho^{4/3}, \tag{4.51}$$

where

$$K_2 = \frac{hc}{8}\left(\frac{3}{\pi}\right)^{1/3}\left(\frac{1+X}{2m_H}\right)^{4/3}. \tag{4.52}$$

Of course there must be a gradual progression between the two formulae (4.49) and (4.51) for intermediate values of $p_0/m_e c$, which is itself determined by the relation:

$$\frac{p_0}{m_e c} = \left(\frac{3h^3 \rho(1+X)}{16\pi m_H m_e{}^3 c^3}\right)^{1/3}. \tag{4.53}$$

Similarly there is not a sharp transition between the perfect gas formula (4.42) and the formulae (4.49) and (4.51); there is a region of temperature and density in which some intermediate and much more complicated formula must be used. When the pressure of a gas is given by equation (4.49) it is said to be *non-relativistically degenerate*; when it is given by equation (4.51) it is said to be *relativistically degenerate*.

In addition to the pressure of the particles, we must consider the radiation pressure. The expression

$$P_{rad} = \frac{1}{3}aT^4 \tag{3.27}$$

is a valid expression for the radiation pressure provided that the radiation is essentially distributed according to the Planck formula. Although this is usually true, there are conditions in stellar interiors when the distribution of radiation with frequency departs seriously from the Planck law. Under these conditions (3.27) may not be a good approximation to the radiation pressure. It appears, however, that almost always, when the formula (3.27) for radiation pressure is not a good approximation, the gas pressure is very much larger than the radiation pressure and the precise value of the radiation pressure is unimportant. Thus we shall assume that (3.27) is always the correct expression for the radiation pressure.

For a fully ionized gas of a given chemical composition, it is possible to divide the $\log \rho - \log T$ plane into regions in which radiation pressure is more important than gas pressure and in which gas pressure is more important than radiation pressure. The latter region can be subdivided into regions where the simple perfect gas law (4.42) holds, where the non-relativistic degeneracy formula (4.49) is a good approximation and where the relativistic degeneracy formula (4.51) holds. These regions are shown in fig. 45.

One of the main reasons for interest in the fact that the stellar gas becomes degenerate at high enough densities arises because of the consequences this has for the Virial Theorem:

$$3\int(P/\rho)\,\mathrm{d}M+\Omega = 0. \tag{3.24}$$

We have previously used this equation to show that any contraction of a star made of a perfect gas leads to its heating up. This arises because, as a star contracts, Ω must become more negative and thus the average value of P/ρ must increase. For a perfect gas of constant mean molecular weight μ this implies that T must increase. Once the stellar

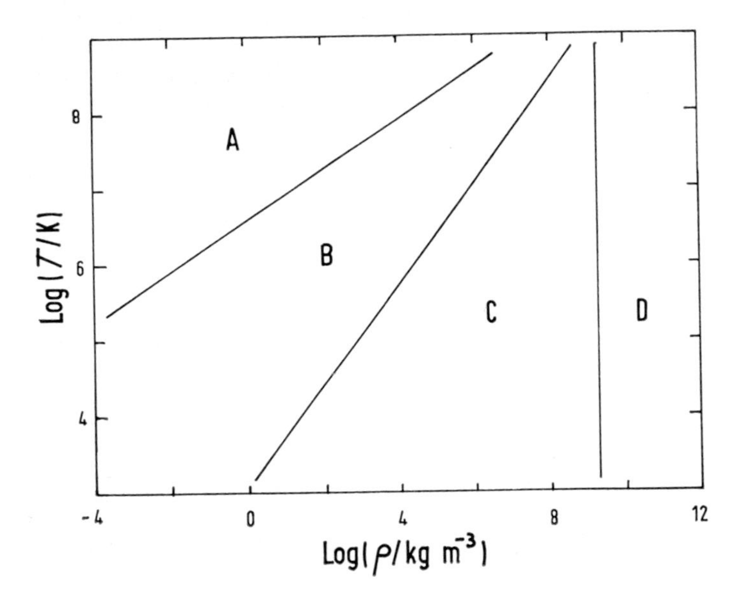

Fig. 45. Pressure as a function of temperature and density. In region A radiation pressure is larger than gas pressure, in region B the material behaves as a perfect gas, in region C the non-relativistic degenerate law holds and in region D the relativistic degenerate approximation is valid.

material is ionized any change in μ is likely to be an increase due to nuclear fusion reactions and this makes it even more definite that the temperature will increase.

Once the stellar gas becomes degenerate, it is possible for P/ρ to increase at the same time as T decreases, because to a first approximation P/ρ given by formulae (4.49) and (4.51) increases as ρ increases and is independent of T. This has the consequence that once the central regions of a star become degenerate they may reach a maximum temperature and then begin to cool down. If this happens no further nuclear fusion reactions will occur in the star and its luminosity will decrease as it contracts until it finally becomes invisible. We have not shown that this will definitely happen, but it now appears that there is a chance that the central temperatures of stars will not rise irrevocably as they evolve. We have already seen in Chapter 2 that the white dwarfs are faint stars of very high density and it is believed that they are cooling degenerate stars and that they represent the final stages of stellar evolution. They will be discussed in Chapter 8.

Summary of Chapter 4

In this chapter we have discussed how pressure, opacity and rate of energy generation depend on temperature, density and chemical composition.

Energy can be released by nuclear fusion reactions building light elements up to nuclei around iron, but most of the energy is released by the conversion of hydrogen into helium. Because positively charged nuclei repel one another, they must have high velocities to approach close enough for short-range nuclear forces to cause a nuclear reaction. Thus nuclear reactions in significant numbers only occur when stellar temperatures are high. The rates of nuclear reactions depend on a high power of the temperature and reactions involving light elements occur at lower temperatures than those involving heavier elements. As a star's central temperature rises while it remains a perfect gas, a succession of nuclear reactions can be expected to occur, with the most important hydrogen burning reactions being the first.

The opacity of stellar material is determined by all the processes which scatter and absorb photons. These include the scattering of radiation by electrons, and the absorption of photons by an atom which causes either a bound electron to move to another bound orbit, or to escape, or an electron to move from one free orbit to another of higher energy. The calculation of opacity requires values for the rates at which many of these processes occur.

Stellar material often behaves as a perfect gas and the value of its pressure can be calculated from Boyle's law. When its density becomes high, the electrons may become so close together that the Pauli exclusion principle restricts their possible momenta. When this is so, the material ceases to be a perfect gas and becomes a degenerate gas. In a highly degenerate gas, the pressure depends only on density and chemical composition and it is then possible for a star to cool down as it loses energy.

Completely accurate expressions for the opacity, rate of energy generation and pressure are extremely complicated, but in some ranges of temperature and density it is possible to find simple mathematical expressions which are good approximations to the physical quantities. These approximate expressions are used in Chapter 5 to obtain qualitative information about the structure of stars.

CHAPTER 5

the structure of main sequence stars

Introduction

IN Chapter 2 we have seen that the majority of stars are main sequence stars and we have already suggested that this could mean either that most stars are main sequence stars for all of their lives or that all stars spend a considerable fraction of their life in the main sequence state. We now believe that the latter is true and that the main sequence phase is one in which stars are obtaining their energy from the conversion of hydrogen into helium which, as we have seen in the last chapter, releases 83% of the maximum energy which can be obtained from nuclear fusion reactions. We also believe that main sequence stars are chemically homogeneous, which means that there are no significant variations of chemical composition from place to place within the stars. As hydrogen burning is the first important nuclear reaction to occur as the central temperature of a star rises, stars should be chemically homogeneous when they reach the main sequence, provided the same was true of the interstellar cloud out of which they were formed. In this chapter we consider the structure of such stars. In the last two chapters we have discussed all of the relevant equations and have concluded that in general their solution can only be obtained with the use of a large computer. However, it is possible to obtain some general properties of chemically homogeneous stars without solving the equations and, in particular, a qualitative understanding of the existence and position of the main sequence in the HR diagram (fig. 46) and of the mass-luminosity relation can be obtained.

In the main part of this chapter we shall discuss the properties of main sequence stars without asking how the stars reach the main sequence and whether the main sequence properties depend on their previous life history. In fact, we believe that for most stars main sequence stellar structure is almost independent of previous life history. Provided no significant nuclear reactions occur before the main sequence is reached, the stars on the main sequence have the chemical composition with which they were formed from the interstellar gas. As the properties of stars change very slowly in the main sequence phase, they can be studied by solving a set of equations in which time dependence does not explicitly enter and which therefore make no reference to the previous history of the star. It is fortunate that main sequence stellar structure is almost independent of previous life history because even now the process of star formation is not very well understood. It has proved possible to study post-main-sequence evolution at a time when very little was known about pre-main-sequence evolution. Some remarks about pre-main-sequence evolution will be made at the end of this chapter.

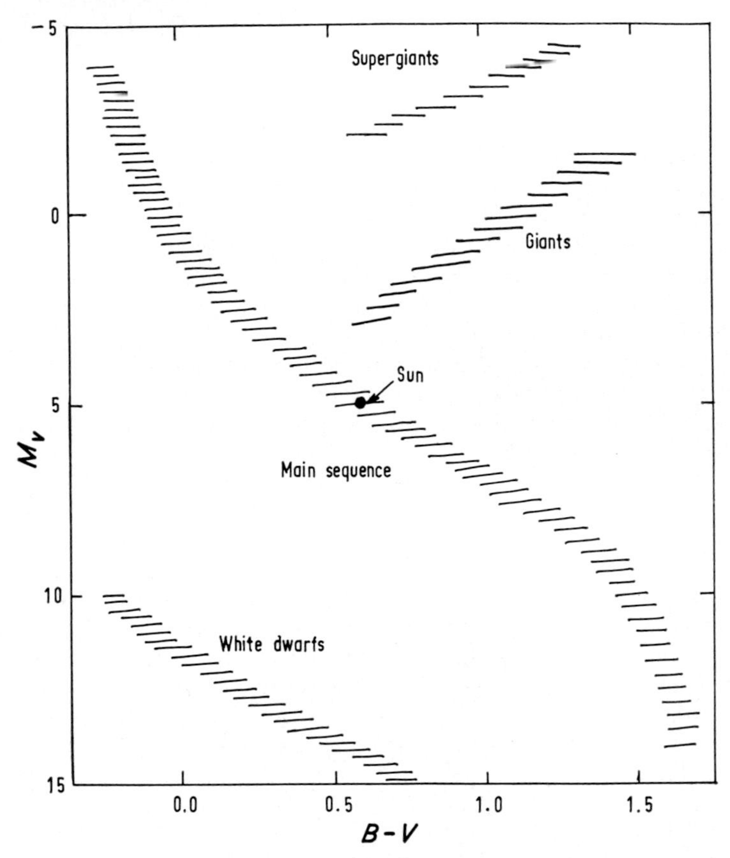

Fig. 46. The Hertzsprung–Russell diagram for nearby stars.

The structure of chemically homogeneous stars

In this chapter we first study the structure of stars which have the same chemical composition, but which differ in mass. Subsequently we shall see how the properties of stars depend on their chemical composition. The equations of stellar structure, which we have discussed in Chapters 3 and 4, are too complicated for us to hope to find an exact analytical solution of them and they must be solved by use of a computer However, provided certain approximations are made, it is possible to discover how properties of a star such as luminosity, radius and effective temperature change, if the mass of the star under consideration is altered, without solving the equations completely. Although this procedure is not a substitute for a complete solution of the problem, the results obtained provide a very useful check on more detailed computations.

The approximations are concerned with the three equations:

$$P = P(\rho, T, \text{composition}), \tag{3.67}$$

$$\kappa = \kappa(\rho, T, \text{composition}) \tag{3.68}$$

and

$$\varepsilon = \varepsilon(\rho, T, \text{composition}), \tag{3.69}$$

which relate the pressure, opacity and rate of energy generation to the density, temperature and chemical composition of the stellar material. We suppose that radiation pressure can be neglected and that the stellar material behaves as a perfect gas so that an expression for the pressure is the perfect gas law:

$$P = \mathscr{R}\rho T/\mu. \tag{5.1}$$

In Chapter 4 we saw that, for appropriate ranges of temperature and density, the laws of opacity and energy generation can be approximately represented by power laws:

$$\kappa = \kappa_0 \rho^{\lambda-1}/T^{\nu-3} \tag{4.39}$$

and

$$\varepsilon = \varepsilon_0 \rho T^{\eta}, \tag{4.23}$$

where λ, ν and η are constants and κ_0 and ε_0 are constant for a given chemical composition. In this chapter we first assume that equations (5.1), (4.39) and (4.23) are accurately true. We also suppose in the first instance that no energy is carried by convection and that, in addition, we may use the simplest boundary conditions discussed in Chapter 3.

$$r = 0, \quad L = 0 \quad \text{at} \quad M = 0 \tag{3.76}$$

and

$$\rho = 0, \quad T = 0 \quad \text{at} \quad M = M_s, \tag{3.77}$$

where M_s is the total mass of the star being considered.

The equations of stellar structure

$$\frac{dP}{dM} = -\frac{GM}{4\pi r^4}, \tag{3.72}$$

$$\frac{dr}{dM} = \frac{1}{4\pi r^2 \rho}, \tag{3.73}$$

$$\frac{dL}{dM} = \varepsilon \tag{3.74}$$

and

$$\frac{dT}{dM} = -\frac{3\kappa L}{64\pi^2 acr^4 T^3} \tag{3.75}$$

must now be solved with the boundary conditions (3.76) and (3.77) and the supplementary relations (5.1), (4.39) and (4.23).

For a sequence of stars of the same homogeneous chemical composition, for which κ_0, ε_0 and μ are the same for all of the stars, we can now show that the properties of a star of any mass can be deduced once the properties of a star of any one mass are known. What we shall in fact show is that the way in which any physical quantity such as luminosity varies from the centre of the star to the surface is the same for stars of all masses and only the absolute value of the luminosity varies from star to star. This is illustrated schematically in fig. 47 where the ratio of luminosity to surface luminosity (L/L_s) is plotted against the *fractional mass*:

$$m \equiv M/M_s. \tag{5.2}$$

We shall show that the curve shown in fig. 47 is the same for all stars with the same laws of opacity and energy generation, but that the value

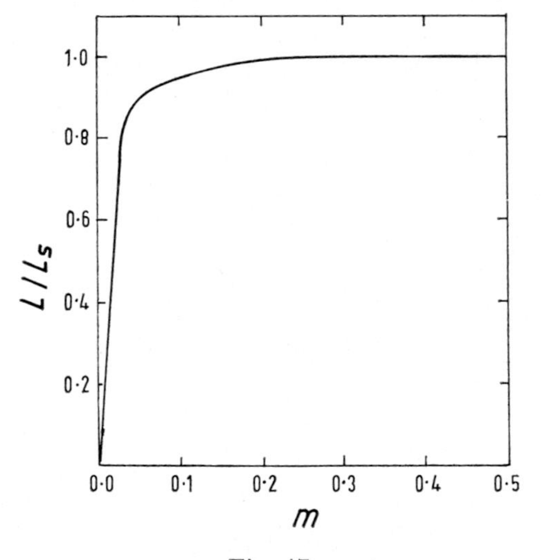

Fig. 47.

of L_s depends on M_s and that it is proportional to some power of M_s, which depends on the values of λ, ν and η in equations (4.39) and (4.23). The same is also true of other quantities such as radius (r_s), effective temperature (T_e) and central temperature (T_c). The equations of stellar structure need only be solved once and the properties of stars of all masses can then be deduced.

Our statement above about the luminosity is equivalent to saying that the luminosity at any point inside a star depends on some power of M_s but otherwise only on the fractional mass m. Mathematically this (and similar statements for the other physical quantities) implies that

$$
\left.\begin{aligned}
r &= M_s^{a_1}\bar{r}(m), \\
\rho &= M_s^{a_2}\bar{\rho}(m), \\
L &= M_s^{a_3}\bar{L}(m), \\
T &= M_s^{a_4}\bar{T}(m) \\
\text{and} \qquad & \\
P &= M_s^{a_5}\bar{P}(m),
\end{aligned}\right\} \tag{5.3}
$$

where a_1, a_2, a_3, a_4 and a_5 are constants and where (as indicated) $\bar{r}$, $\bar{\rho}$, $\bar{L}$, $\bar{T}$ and $\bar{P}$ depend only on the fractional mass m. We now verify that expressions of the form of (5.3) do satisfy the equations of stellar structure provided that values of the constants a_1 to a_5 are chosen correctly.

We consider each equation in turn. Equation (3.72) can be written:

$$M_s^{a_5-1}\mathrm{d}\bar{P}/\mathrm{d}m = -GM_s^{1-4a_1}m/4\pi\bar{r}^4.$$

If this equation is to be true for all values of M_s, the powers of M_s entering on the two sides of the equation must be the same. Thus:

$$4a_1 + a_5 = 2, \tag{5.4}$$

and the equation becomes:

$$\mathrm{d}\bar{P}/\mathrm{d}m = -Gm/4\pi\bar{r}^4. \tag{5.5}$$

Similarly equation (3.73) can be written:

$$M_s^{a_1-1}\mathrm{d}\bar{r}/\mathrm{d}m = 1/4\pi M_s^{2a_1+a_2}\bar{r}^2\bar{\rho},$$

which will reduce to an equation independent of M_s if

$$3a_1 + a_2 = 1, \tag{5.6}$$

and which then has the form:

$$\mathrm{d}\bar{r}/\mathrm{d}m = 1/4\pi\bar{r}^2\bar{\rho}. \tag{5.7}$$

Before separating (3.74) into an algebraic equation and a differential equation, the expression (4.23) for ε can be substituted on the right-hand side. Then the equation can be written:

$$M_s^{a_3-1}\mathrm{d}\bar{L}/\mathrm{d}m = \varepsilon_0 M_s^{a_2+\eta a_4}\bar{\rho}\bar{T}^\eta.$$

This is independent of M_s if

$$a_3 = 1 + a_2 + \eta a_4, \tag{5.8}$$

and then

$$\mathrm{d}\bar{L}/\mathrm{d}m = \varepsilon_0\bar{\rho}\bar{T}^\eta. \tag{5.9}$$

If expression (4.39) for κ is substituted into equation (3.75), this can be written:

$$M_s^{(a_4-1)}\mathrm{d}\bar{T}/\mathrm{d}m = -3\kappa_0 M_s^{[a_3+(\lambda-1)a_2-4a_1-\nu a_4]}\bar{\rho}^{(\lambda-1)}\bar{L}/64\pi^2ac\bar{r}^4\bar{T}^\nu.$$

This is independent of M_s if

$$4a_1+(\nu+1)a_4 = (\lambda-1)a_2+a_3+1, \quad (5.10)$$

and then

$$\mathrm{d}\bar{T}/\mathrm{d}m = -3\kappa_0\bar{\rho}^{(\lambda-1)}\bar{L}/64\pi^2ac\bar{r}^4\bar{T}^\nu. \quad (5.11)$$

Finally equation (5.1) gives:

$$a_5 = a_2+a_4 \quad (5.12)$$

and

$$\bar{P} = \mathscr{R}\bar{\rho}\bar{T}/\mu. \quad (5.13)$$

We now have five algebraic equations (5.4), (5.6), (5.8), (5.10) and (5.12) for the constants $a_1 \ldots a_5$. These are *inhomogeneous algebraic equations* which means that some of them contain terms which are independent of the a's. Such a system of inhomogeneous equations will have consistent solutions provided that the determinant of the coefficients of $a_1 \ldots a_5$ does *not* vanish. The condition for the determinant to vanish is that

$$\nu-3\lambda = \eta+3, \quad (5.14)$$

and this is not true for any of the approximate laws of opacity and energy generation which we have considered in Chapter 4; η is usually large and positive while $\nu-3\lambda$ is close to zero. Thus it appears that equations (5.4), (5.6), (5.8), (5.10) and (5.12) can be solved to give unique values for the constants $a_1 \ldots a_5$. Because the general solution is complicated it will not be written down, but solutions will shortly be given for special values of λ, ν and η.

To obtain the details of the structure of a star of any given mass the differential equations (5.5), (5.7), (5.9) and (5.11) and the equation (5.13) must now be solved to find $\bar{r}$, $\bar{\rho}$, $\bar{L}$, $\bar{T}$ and $\bar{P}$ in terms of m. The centre and surface of the star are $m = 0$ and $m = 1$ respectively and the boundary conditions are

$$\bar{r} = \bar{L} = 0 \quad \text{at} \quad m = 0 \quad (5.15)$$

and

$$\bar{\rho} = \bar{T} = 0 \quad \text{at} \quad m = 1. \quad (5.16)$$

This set of equations can now be solved on a computer and after the solution has been obtained the quantities $\bar{r}$, $\bar{\rho}$, etc. can be converted into r, ρ, etc. for a star of any given mass M_s by using the relations (5.3) and the values of the constants $a_1 \ldots a_5$ previously found. As mentioned earlier, the equations have only to be solved once and the properties of stars of all masses can then be obtained. Such a set of models of stars in which the dependence of the physical quantities on fractional mass m is independent of the total mass of the star is known as a *homologous sequence of stellar models*.

Mass–luminosity and luminosity–effective temperature relations

For such homologous stellar models there is clearly a mass–luminosity relation and also a simple relation between luminosity and effective temperature such as that which characterizes the main sequence in the HR diagram. Thus we have shown that at any point inside such a star:

$$L = M_s^{a_3}\bar{L}(m).$$

At the surface of the star ($m = 1$) this equation becomes:

$$L_s = M_s^{a_3}\bar{L}(1). \tag{5.17}$$

Since $\bar{L}(1)$ is the same for all stars of the same chemical composition, the luminosity should be proportional to the a_3th power of the mass. Values of a_3 will be considered shortly and they will be found to give a mass–luminosity relation similar to that found observationally for main sequence stars.

In addition

$$r_s = M_s^{a_1}\bar{r}(1), \tag{5.18}$$

while

$$L_s = \pi ac r_s^2 T_e^4. \tag{2.7}$$

Combining equations (5.17), (5.18) and (2.7) it can be seen that

$$T_e = M_s^{(a_3 - 2a_1)/4}[\bar{L}(1)/\pi ac\bar{r}^2(1)]^{1/4}. \tag{5.19}$$

Equations (5.17) and (5.19) then show that for the homologous sequence of stars:

$$L_s \propto T_e^{4a_3/(a_3 - 2a_1)}. \tag{5.20}$$

This shows that the stars lie on a straight line in the theoretical HR diagram (plot of $\log L_s$ against $\log T_e$) and this might be identified with the main sequence.

Solutions for particular laws of opacity and energy generation

Two things must now be considered. Do the values of λ, ν and η, which have been suggested in Chapter 4, lead to a theoretical mass–luminosity relation and main sequence which are in reasonable agreement with the observations, and to what extent are the various assumptions which led to the homologous sequence of models valid? We first study the homologous solutions in more detail and then discuss their limitations.

In Chapter 4 two particular approximations to the opacity law which have been mentioned are (equations (4.36) and (4.37)):

$$\kappa = \kappa_1. \qquad \lambda = 1,\ \nu = 3 \tag{5.21}$$

and

$$\kappa = \kappa_2\rho/T^{3.5}. \quad \lambda = 2,\ \nu = 6{\cdot}5. \tag{5.22}$$

Reasonable approximations to the rate of energy generation by the proton–proton chain and the carbon–nitrogen cycle are (see equations (4.21) and (4.22)):

$$\varepsilon = \varepsilon_0 \rho T^4, \qquad \eta = 4 \tag{5.23}$$

and

$$\varepsilon = \varepsilon_0 \rho T^{17}, \qquad \eta = 17. \tag{5.24}$$

For the four possible combinations of these laws of opacity and energy generation the constants $a_1 \ldots a_5$ have been calculated and they are shown in Table 4. In addition, the quantity $4a_3/(a_3 - 2a_1)$ which enters into the relation between luminosity and effective temperature is tabulated and it is denoted by a_6.

λ	ν	η	a_1	a_2	a_3	a_4	a_5	a_6
1	3	4	3/7	−2/7	3	4/7	2/7	28/5
1	3	17	4/5	−7/5	3	1/5	−6/5	60/7
2	6·5	4	1/13	10/13	71/13	12/13	22/13	284/69
2	6·5	17	9/13	−14/13	67/13	4/13	−10/13	268/49

Table 4. Constants occurring in homology relations for four laws of opacity and energy generation.

It should be stressed that it is quite possible that not all of these combinations of laws of opacity and energy generation will occur in stars. Each approximation is valid in some range of temperature and density as we have already seen in Chapter 4, but we do not know in advance what values of temperature and density will occur in a star of a given mass. If we calculate a series of models of stars of different masses using an opacity law (5.22) and law of energy generation (5.23), we must decide afterwards whether the physical conditions in the models are such that these laws would be valid. What we expect to find is that for stars in a certain mass range the results will be consistent; the physical conditions in the stars will be such that the laws of opacity and energy generation assumed are a good approximation. For stars outside that mass range another approximation to the laws would have been more appropriate and the calculations must be repeated.

What can be said straight away from the results of Table 4 is as follows. All of the opacity and energy generation laws predict a mass–luminosity relation with the luminosity proportional to a power of the mass between 3 and 5·5. The observational mass–luminosity relation discussed in Chapter 2 is not a simple power law but, if approximated to by a power law, it has an exponent in the same range. Thus the observations suggest that for main sequence stars of about solar mass $L_s \propto M_s^5$ while for more massive stars $L_s \propto M_s^3$. In addition it may be noted that the exponent depends strongly on the law of opacity but only slightly on the law of energy generation.

In fact, the mass–luminosity relation was reasonably well understood before the law of energy generation in stars was known accurately and even before the nuclear origin of stellar energy had been established. The reason for this is as follows. If we assume some power law

$$\varepsilon = \varepsilon_0 \rho^{\alpha} T^{\eta} \tag{5.25}$$

for the law of energy generation, but do not know the values of α and η, homologous solutions of the equations certainly exist and a relation can be obtained between the constants a_3 and a_1 in equation (5.3) which does not involve α and η. Thus equations (5.4), (5.6), (5.10) and (5.12) combine to give:

$$a_3 = (\nu - \lambda + 1) + (3\lambda - \nu) a_1, \tag{5.26}$$

and, if this is combined with equation (5.3) evaluated at $m = 1$, it can be seen that

$$L_s = [\bar{L}(1)/\bar{r}(1)^{3\lambda - \nu}] M_s^{\nu - \lambda + 1} r_s^{3\lambda - \nu}. \tag{5.27}$$

Equation (5.27) is a mass–luminosity–radius relation which can only be converted to a mass–luminosity relation when the law of energy generation is known so that the modified equation (5.8), with α included as well as η, can be used to enable a_1 to be eliminated from equation (5.26). However, for the laws of opacity (5.21), (5.22) it can be seen that $3\lambda - \nu$ is zero or small and the dependence of luminosity on radius is slight. Eddington obtained the relation (5.27) before anything was known about nuclear reactions in stars and he showed that it was in qualitative agreement with the observed mass–luminosity relation.

Effect of variation of chemical composition

We have discussed above the properties of stars of different mass and the same chemical composition. It is now possible to ask how the properties of a star of a given mass are altered if its chemical composition is changed. Provided we still assume that radiation pressure can be neglected and that the laws of opacity and energy generation are power laws, it is possible to find how the properties of a star vary with chemical composition without solving the equations completely. We will discuss this for one particular pair of laws of opacity and energy generation but the results are quite typical. The laws chosen are reasonable first approximations for stars of about a solar mass.

Suppose that the law of opacity is:

$$\kappa = \kappa_0 Z(1 + X)\rho / T^{3\cdot 5}, \tag{5.28}$$

and the law of energy generation is:

$$\varepsilon = \varepsilon_0 X^2 \rho T^4, \tag{5.29}$$

where, as in Chapter 4, X and Z are the fractional abundances by mass of hydrogen and the heavy elements and where we have now made explicit the dependence of the laws of opacity and energy generation on the chemical composition of the star. Suppose also that Z is so small that a good approximation to the mean molecular weight is:

$$\mu = 4/(3+5X). \tag{4.47}$$

It can now be shown that, with the laws (5.28), (5.29), equations (3.72) (3.75) and (5.1) have solutions of the form:

$$\left.\begin{aligned} r &= r_1(X)r_2(Z)r_3(M), \\ \rho &= \rho_1(X)\rho_2(Z)\rho_3(M), \\ L &= L_1(X)L_2(Z)L_3(M), \\ T &= T_1(X)T_2(Z)T_3(M) \\ \text{and} \qquad & \\ P &= P_1(X)P_2(Z)P_3(M). \end{aligned}\right\} \tag{5.30}$$

If this is true it follows that, if we change the chemical composition of a star of a given mass, we can predict how its properties such as luminosity, radius and effective temperature will change without solving the full equations.

The detailed argument is rather similar to that given for the case of stars of the same chemical composition and varying mass and it will not be given here. The key factor in the solution is that X and Z only occur algebraically in the equations and algebraic expressions for $r_1(X)$, $r_2(Z)$, etc. must be found so that the terms in X and Z are the same on both sides of all the equations. The explicit form of the X and Z dependence of the solutions (5.30) is given below and it can be verified that, if these expressions are substituted into equations (3.72) to (3.75) and (5.1) they do lead to equations for r_3, ρ_3, L_3, T_3 and P_3 in terms of M in which X and Z do not occur. The solutions are:

$$\left.\begin{aligned} r &= X^{4/13}(1+X)^{2/13}(3+5X)^{7/13}Z^{2/13}r_3(M), \\ \rho &= X^{-12/13}(1+X)^{-6/13}(3+5X)^{-21/13}Z^{-6/13}\rho_3(M), \\ L &= X^{-2/13}(1+X)^{-14/13}(3+5X)^{-101/13}Z^{-14/13}L_3(M), \\ T &= X^{-4/13}(1+X)^{-2/13}(3+5X)^{-20/13}Z^{-2/13}T_3(M) \\ \text{and} \qquad & \\ P &= X^{-16/13}(1+X)^{-8/13}(3+5X)^{-28/13}Z^{-8/13}P_3(M). \end{aligned}\right\} \tag{5.31}$$

It is also possible to combine the expressions for L and r evaluated at the surface $M = M_s$ to obtain an expression for how the effective temperature depends on the abundance of hydrogen and the heavy elements. This expression is:

$$T_e = X^{-5/26}(1+X)^{-9/26}(3+5X)^{-115/52}Z^{-9/26}[L_3(M_s)/\pi acr_3{}^2(M_s)]^{1/4}. \tag{5.32}$$

From the results (5.31), (5.32) it can be seen that if either X is increased keeping Z constant, or if Z is increased keeping X constant, or if both are increased, there is an increase in the radius of the star and a decrease in both the luminosity and the effective temperature. The changes refer to a star of a given mass. If the position of the main sequence in the HR diagram is considered it is found that when the composition is changed the displacement of the main sequence is smaller than the change in position of individual stars. This is illustrated in fig. 48. In this diagram are shown main sequences for three different chemical compositions and the positions on each of the sequences of a star of a given mass. This means that homogeneous stars of different chemical compositions lie on a rather broadened main sequence.

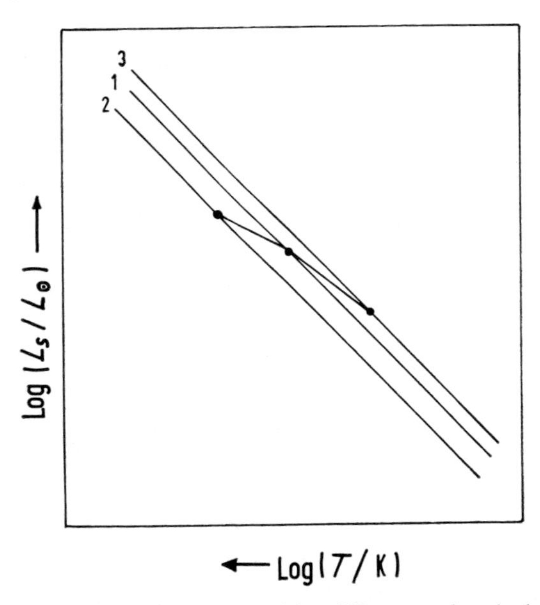

Fig. 48. Three main sequences with different chemical compositions. Sequence 2 contains more helium and sequence 3 more heavy elements than sequence 1. The filled circles show the three positions of a star of given mass.

One result of the present calculations, which proves to be of great importance when it is confirmed by calculations in which more accurate laws of opacity and energy generation are used, is that stars with a homogeneous chemical composition which are releasing energy through nuclear reactions are situated in the general region of the observed main sequence and they do not lie in the region of the giants, supergiants and white dwarfs. The position of a star in the main sequence band does depend on its chemical composition, but for all possible chemical compositions there are the important regions in the HR diagram mentioned above which are not covered. This means, for example, that, if stars are well mixed as they convert hydrogen into helium so that their chemical composition remains uniform, they remain in the neighbourhood of the main sequence as they evolve. The same may not be true if, as was generally assumed in Chapter 3, the changes of chemical composition are localized where they occur so that the star becomes chemically non-uniform and this will be discussed further in the next chapter.

All of these results suggest that it is plausible that main sequence stars are in fact stars of homogeneous chemical composition which are gradually burning hydrogen into helium in their interiors and that the precise position of a star on the main sequence is determined primarily by its mass and to some extent by its chemical composition. The discussion given above is, however, far from conclusive as many approximations have been made in discussing these homologous models. The approximations may be listed again as follows:

(i) Neglect of convection.
(ii) Neglect of radiation pressure.
(iii) Use of simple formulae for ε and κ.
(iv) Use of the boundary condition $\rho = T = 0$ at $M = M_s$ instead of a more realistic condition.

The existence of the homologous models depended crucially on the expressions (4.39), (4.23) and (5.1) for κ, ε and P being products of powers of ρ and T, and on the boundary conditions having simple expressions at $m = 0$ and $m = 1$. Any modification of assumptions (ii), (iii) and (iv) will almost certainly mean that such homologous solutions do not exist. Thus, for example, if equation (5.1) is replaced by:

$$P = \frac{\mathscr{R}\rho T}{\mu} + \frac{1}{3}aT^4,$$

all the terms in this equation will only depend on the same power of M_s if the values of λ, ν and η are such that $a_2 = 3a_4$.

If (ii), (iii) and (iv) are assumed to be valid and convection is supposed to occur only in an inner region in which all of the energy generation occurs and in which convection is so efficient that

$$\frac{P}{T}\frac{\mathrm{d}T}{\mathrm{d}P} = \frac{\gamma - 1}{\gamma}, \tag{3.86}$$

then homologous solutions can still be obtained. In the inner region, where (3.86) is valid, this equation takes the form

$$\frac{\bar{P}}{\bar{T}}\frac{\mathrm{d}\bar{T}}{\mathrm{d}\bar{P}} = \frac{\gamma - 1}{\gamma} \tag{5.33}$$

for any values of a_4 and a_5 in equations (5.3), and the values of a_4 and a_5 required in the outer regions of the star can also be used in the convective core. The demonstration that homologous solutions do exist is rather complicated as the inner region in which convection is occurring is surrounded by an outer region in which energy is carried by radiation and only the result can be quoted here. It is found that stars with convective cores and power laws for opacity and energy generation all have the same fraction of their mass in the convective core.

If a star has an outer convective region where, as discussed on page 79, convection is unlikely to be so efficient that (3.86) is valid everywhere, the equations governing the convective region are much more complicated and there are no longer homologous solutions even if (ii), (iii) and (iv) are assumed true. Later in this chapter we will discuss circumstances when assumption (iv) cannot be used and we will then discuss what is a more realistic boundary condition.

We have now given a long enough discussion of the properties of homologous stellar sequences, which are obtained if approximate expressions for the laws of pressure, opacity and energy generation are used. We have probably obtained a qualitative understanding of the properties of homogeneous stars but, if any quantitative comparison is to be made between theory and observation, the best possible expressions must be used for the physical quantities. As the equations of stellar structure can then only be solved on a computer, we can do no more than discuss the results which have been obtained and what, if anything, remains to be done.

General properties of homogeneous stars

The general properties of stars, which are burning hydrogen into helium and which are chemically homogeneous either because the burning has only just started, or because for some reason they are well mixed, are as follows:

(i) High mass stars have a region in the centre in which a significant proportion of the energy is carried by convection. The central temperature is an increasing function of stellar mass, as was suggested by the homology results of Table 4, and this means that in the high mass stars energy generation will be by the CN cycle rather than the PP chain. In addition, as predicted by most of the homologous results, the central density decreases with increasing stellar mass. From fig. 43 in Chapter 4 it can be seen that at sufficiently high temperatures an increase in temperature, or a decrease in density, tends to make the opacity scarcely dependent on temperature and density; as has been stated on page 101, this occurs when electron scattering is the main source of opacity. This means that the second row of Table 4 (constant opacity and CN energy generation) is likely to give the best approximation to the properties of these stars. Convection occurs in the central regions of these stars because the CN cycle provides a highly concentrated source of energy and radiation alone is not adequate to carry the energy away from the central regions. The existence of a convective core in these stars implies that the material in the deep interior is well mixed and this in turn, as we shall see in the next chapter, has important consequences for their evolution.

(ii) In lower mass stars, the main energy generation is by the PP chain and the opacity is more nearly due to Kramers' law:

$$\kappa = \kappa_2 \rho / T^{3\cdot 5}. \tag{5.22}$$

The source of nuclear energy is no longer sufficiently concentrated to give rise to a convective core. These stars do, however, have outer regions in which convection occurs. This can be explained as follows. The surface temperatures of high mass main sequence stars are higher than those of low mass stars as has been predicted by the homologous solutions in Table 4 and they are such that the abundant elements, hydrogen and helium, are ionized at the stellar surface. In the low mass stars the surface temperatures are lower and these elements are neutral. In that case, there is a region just below the stellar surface in which these elements are being ionized and in these ionization zones the ratio of specific heats of the stellar material is very much smaller than usual and can be comparable with unity. As a result the criterion for convection

$$\frac{P}{T}\frac{\mathrm{d}T}{\mathrm{d}P} > \frac{\gamma-1}{\gamma} \tag{3.65}$$

is satisfied. In these outer convection zones the structure of the star does depend on the efficiency with which convection can carry energy and the lack of a really good theory of convection means that there is an important uncertainty in the theory. *This is probably the biggest single uncertainty in the theory of main sequence stars being more important than inaccuracies in the laws of opacity and energy generation.* We can turn the discussion round and ask how much energy must be carried by convection if theory and observation are to agree and the most that can be said is that the quantity does not seem implausible.

(iii) There is a mass below which no consistent solution of the equations can be obtained. As has been mentioned earlier, the central temperature is an increasing function of stellar mass while the opposite is true of the central density. This means that as the stellar mass is reduced, there is an increasing tendency for the material in the centre of the star to become an imperfect or degenerate gas; we have seen in Chapter 4 that the perfect gas law breaks down at low temperature and high density. When the evolution of very low mass stars is studied it is found that, although there is initially an increase in the central temperature, it reaches a maximum value and decreases again when the centre becomes a degenerate gas before any significant nuclear reactions converting hydrogen and helium take place. Such stars have a very brief existence in astronomical terms as they cool down and die without ever *tapping* a nuclear energy source and having a main sequence phase. The critical mass below which this happens depends on the chemical composition of the star, but it is about $0{\cdot}1\ M_{\odot}$ according to recent calculations. It is difficult to observe stars of such low mass, even if they exist, and it is not at present clear that this theoretical prediction of a minimum mass for main sequence stars is confirmed by observation.

The critical mass depends fairly slightly on chemical composition provided that the main sequence stars are obtaining their energy by the burning of hydrogen into helium. If calculations are made for stars which contain no hydrogen and which are mainly composed of helium so that the main source of energy is helium burning, the lowest mass for which a consistent main sequence model can be obtained is about $0{\cdot}35M_{\odot}$; for a pure carbon star it is even higher at about $0{\cdot}8M_{\odot}$. As present evidence suggests that all main sequence stars do contain a considerable proportion of hydrogen, these latter results may be of only theoretical interest unless it is possible for stars to lose all of their hydrogen at a later stage of evolution and become pure helium stars.

(iv) The qualitative results concerning the existence of a main sequence and a mass–luminosity relation and the dependence of the position of the main sequence on chemical composition found for homologous stars are confirmed by more careful calculations. The main sequence is no longer found to be precisely a straight line and the mass–luminosity relation is no longer a pure power law. It is also true that the observed relations between mass and luminosity and luminosity and effective temperature are not simple power laws.

(v) Because the luminosity of a star depends on a fairly high power of its mass ($L_s \propto M_s^3$ or $\propto M_s^5$ depending on mass range) while the nuclear energy which can be released in converting a given fraction of its mass from hydrogen into helium is directly proportional to the mass, the time for which stars can exist in a hydrogen burning phase is smaller for stars of greater mass. Thus the total energy release in converting hydrogen into helium if a star is initially composed of pure hydrogen and all of the hydrogen is burnt is:

$$E_{\mathrm{H}\rightarrow\mathrm{He}} = 0{\cdot}007 M_s c^2. \tag{5.34}$$

This will be an over-estimate of the actual amount of energy released in the hydrogen burning phase both because a star is unlikely to be made of pure hydrogen and because, as we shall see later, 100% efficiency in the conversion of hydrogen to helium is unlikely. If the main sequence luminosity when the star is made of pure hydrogen is L_s, an estimate of the time for which hydrogen burning can occur (the *main sequence lifetime* of the star) is:

$$t_{H \rightarrow He} = 0 \cdot 007 M_s c^2 / L_s. \qquad (5.35)$$

This is likely to be an over-estimate of the time, not only for the two reasons listed above, but also because L_s will increase as the conversion of hydrogen into helium proceeds. We have already seen that equation (5.31) predicts this property for stars which remain homogeneous as they evolve. We shall see in the next chapter that the same is also true for stars which develop non-uniformities of chemical compositions as they evolve.

Estimates of the main sequence lifetimes of stars of different masses, based on recent calculations of the evolution of hydrogen burning stars are shown in Table 5. These times are shorter than

$M/M_\odot$	15·0	9·0	5·0	3·0	2·25	1·5	1·25	1·0
Lifetime	$1{\cdot}0 \times 10^7$	$2{\cdot}2 \times 10^7$	$6{\cdot}8 \times 10^7$	$2{\cdot}3 \times 10^8$	$5{\cdot}0 \times 10^8$	$1{\cdot}7 \times 10^9$	$3{\cdot}0 \times 10^9$	$8{\cdot}2 \times 10^9$

Table 5. Main sequence lifetime (in years) for stars of different masses.

those obtained by a simple use of equation (5.35) because allowance has been made for the factors causing (5.35) to be an over-estimate. One very important result can be deduced from this table. It is known, from the properties of the radioactive elements in the Earth's crust, that the Earth has been solid for about $4{\cdot}5 \times 10^9$ years. From Table 5 it is clear that *main sequence stars with a mass greater than about* $1{\cdot}25\ M_\odot$ *cannot have been on the main sequence when the Earth solidified.*

As the first important release of nuclear energy occurs when stars are on the main sequence, the pre-main-sequence lifetime (which will be discussed briefly later in this chapter) is thought to be governed by the release of gravitational energy giving a lifetime of order 3×10^7 years for the Sun (equation (3.40)) and a lifetime of less than $4{\cdot}5 \times 10^9$ years for all stars in the mass range which is observed. Thus not only can we say that massive stars cannot have been on the main sequence all of the time since the Earth solidified, but that any massive main sequence star which we observe today must have been formed long since the Earth solidified. It can be seen from the Table that stars more massive than $15M_\odot$ have probably been formed in the last ten million years. This is perhaps the first definite evidence that not all astronomical objects came into existence at the same time and it makes it very likely that genuine new stars (as opposed to novae which are old stars becoming brighter) are being formed today.

It should be stressed that these new stars are not being created out of nothing. Most astronomers believe that our Galaxy came into being as a mass of gas between 10^{10} and 2×10^{10} years ago. During the time that has passed since then, stars have formed out of condensations in this gas. What we are saying in this section is that all of the stars were not formed at once, but that we have evidence that there has been a continuing process of star formation throughout the lifetime of the Galaxy. As there is still interstellar gas and dust in the Galaxy, raw material is available for the formation of new stars today. From Table 5 it can be deduced that stars of one solar mass and less formed quite early in the lifetime of the Galaxy are still luminous today whereas massive stars formed early in the galactic lifetime must have completed their life history long ago. Whenever a star loses mass to the interstellar medium, as occurs in the explosion of novae and supernovae, this material can then be incorporated in a future generation of stars. Some star clusters, whose HR diagrams have been discussed in Chapter 2, contain very luminous, and hence presumably massive, stars on their main sequences. If the idea expressed in that chapter that all stars in a cluster have essentially the same age is true, this means that *entire clusters of stars have been formed quite recently in the lifetime of the Galaxy*. This will be discussed further in the following chapter.

(vi) As mentioned earlier, whatever chemical composition is assumed for the stars, there are regions in the HR diagram which are not covered by the set of possible main sequences. *Giants, supergiants and white dwarfs cannot be chemically homogeneous stars in which important nuclear energy release is taking place.* We believe that giants and supergiants have important non-uniformities of chemical composition and that there is no significant nuclear energy release in white dwarfs. The structure of giants and supergiants will be discussed in Chapter 6 and that of white dwarfs in Chapter 8.

Detailed calculations of main sequence structure and comparison with observation

In any really detailed comparison of the results of this section with observation there are a variety of difficulties. There are inaccuracies in the theoretical calculations because of uncertainties in such quantities as the opacity, energy generation and transport of energy by convection. There are the problems discussed in Chapter 2 of converting the observed magnitudes and colour indices into the bolometric magnitude and effective temperature calculated by the theorist. Quantities such as mass and radius can only be observed directly for a limited number of stars. Finally, only the chemical composition of the outer layers of a star can be deduced from observations and there is no direct evidence about whether or not a star has a homogeneous chemical composition. For these reasons the results obtained by different authors do not agree in fine detail even though we believe that the main sequence phase of stellar evolution is generally well understood. Some recent results are described below.

The main sequences for four different chemical compositions which are within the range observed in stellar surfaces, are shown below in

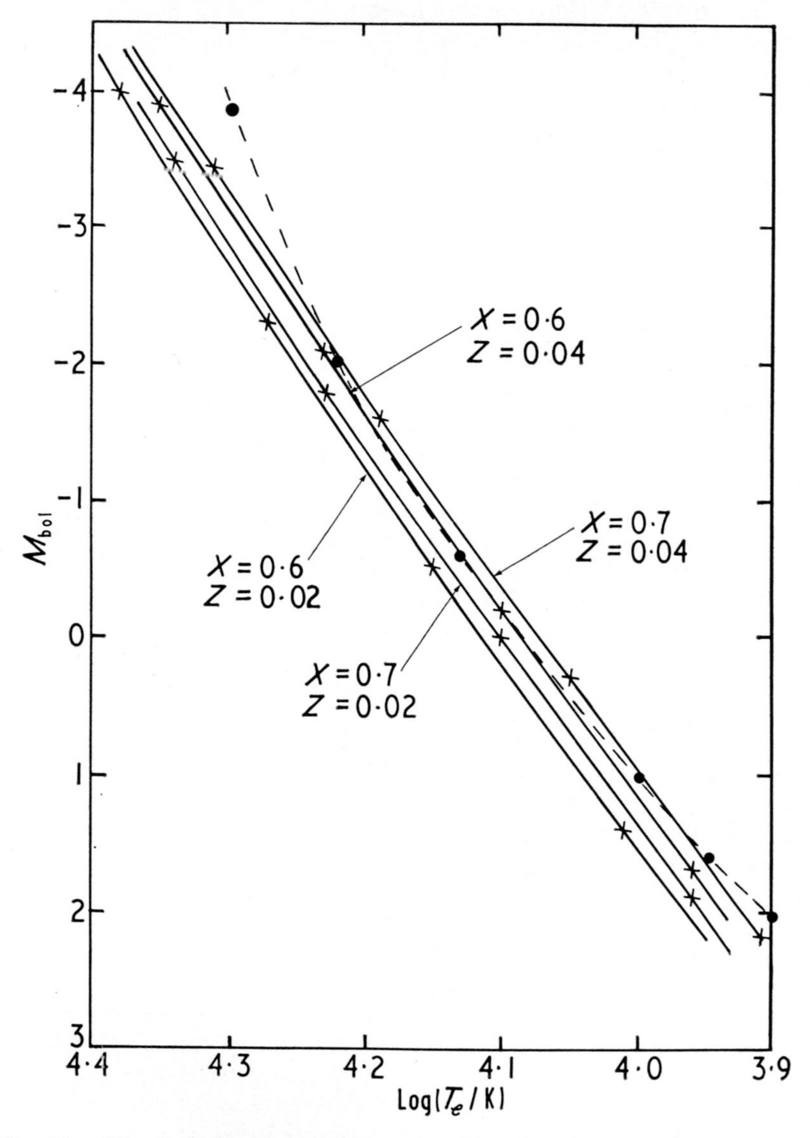

Fig. 49. Theoretical main sequences for four chemical compositions. Also shown as a dashed curve is a section of the observed main sequence.

fig. 49. The compositions are:

$$\left.\begin{array}{llll} (a) & X = 0{\cdot}60, & Y = 0{\cdot}38, & Z = 0{\cdot}02, \\ (b) & X = 0{\cdot}70, & Y = 0{\cdot}28, & Z = 0{\cdot}02, \\ (c) & X = 0{\cdot}60, & Y = 0{\cdot}36, & Z = 0{\cdot}04, \\ (d) & X = 0{\cdot}70, & Y = 0{\cdot}26, & Z = 0{\cdot}04. \end{array}\right\} \quad (5.36)$$

It can be seen that the calculated main sequences are approximately straight lines with a slope in general agreement with the observed main sequence band. Also shown in the diagram is a mean line through the main sequence deduced from observations of nearby stars with appropriate conversion from observed quantities to $\log L_s$ and $\log T_e$. Of course the observed main sequence band is quite broad and all of the above lines are within the region of the observed main sequence. It can be noted from fig. 49 that the position of the main sequence is particularly sensitive to a change in the heavy element content, Z, and this is in general agreement with the homology result of equation (5.32).

For a slightly different chemical composition than any of the above:

$$X = 0{\cdot}71, \quad Y = 0{\cdot}27, \quad Z = 0{\cdot}02, \quad (5.37)$$

I. Iben has recently calculated main sequence models for a variety of masses between $0{\cdot}5\ M_\odot$ and $15 M_\odot$ and he has also studied both pre-

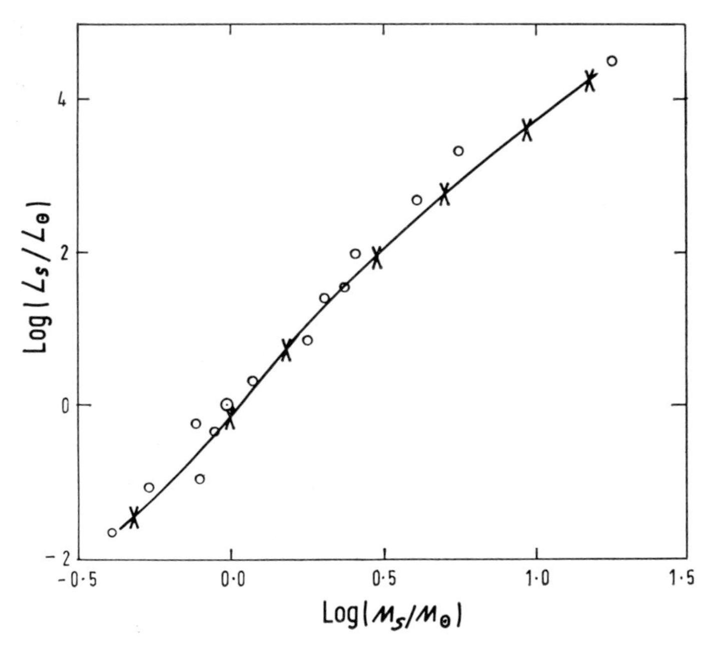

Fig. 50. A theoretical mass–luminosity relation based on calculations by Iben. Also shown are some stars of well-determined mass and luminosity including the Sun, ⊙.

and post-main-sequence evolution of these stars. From his main sequence results it is possible to draw up a theoretical main sequence mass–luminosity relation and that is shown in fig. 50. Also shown in the diagram are points corresponding to nearby stars of known mass and luminosity and it can be seen that once again there is a good qualitative agreement between theory and observation. Table 6 shows what fraction of the mass of each star is in either a convective core or a convective envelope.

$M/M_\odot$	15·0	9·0	5·0	3·0	1·5	1·0	0·5
M_{cc}	0·38	0·26	0·21	0·17	0·06	0·00	0·01
M_{ce}	0·00	0·00	0·00	0·00	0·00	0·01	0·42

Table 6. Fractional mass in convective core (M_{cc}) or convective envelope (M_{ce}) for main sequence stars of different masses.

It was mentioned in Chapter 2 that there is a group of stars known as sub-dwarfs which lie below the main sequence in the HR diagram. It was also mentioned that the sub-dwarfs appeared to have a low content of heavy elements. It can be seen from the results given above that stars with low Z do lie below stars with high Z and the low content of heavy elements may go at least some way towards explaining the position of the sub-dwarfs in the HR diagram.

Pre-main-sequence evolution

We will now give a brief description of what is at present known about pre-main-sequence evolution. As mentioned in the introduction to this chapter, the subject of star formation and pre-main-sequence evolution is at present not very well understood and much of what will be said in the remainder of this chapter would be disputed by some workers in the field. In the first instance if, as we believe, stars have been formed out of condensations in the interstellar gas, their initial state must be one of very large radius and very low luminosity and surface temperature; that is, a newly formed star is low down to the right in the HR diagram. When the pre-main-sequence evolution of stars was first discussed it was assumed that the evolution to the main sequence was a process in which the star's radius decreased and the luminosity and surface temperature increased steadily. Calculations in which it was assumed that all of the energy transport in protostars was by radiation gave results with essentially that character and these are shown in fig. 51. The surface temperature increases steadily and the radius decreases steadily but there is a maximum in the luminosity just before the main sequence is reached. This occurs because, when nuclear reactions commence in the central regions of the stars, they lead to a rise in temperature and pressure in the central regions. This is followed by a small expansion in the central regions and a decrease in temperature and luminosity.

When the properties of these contracting protostars were studied in

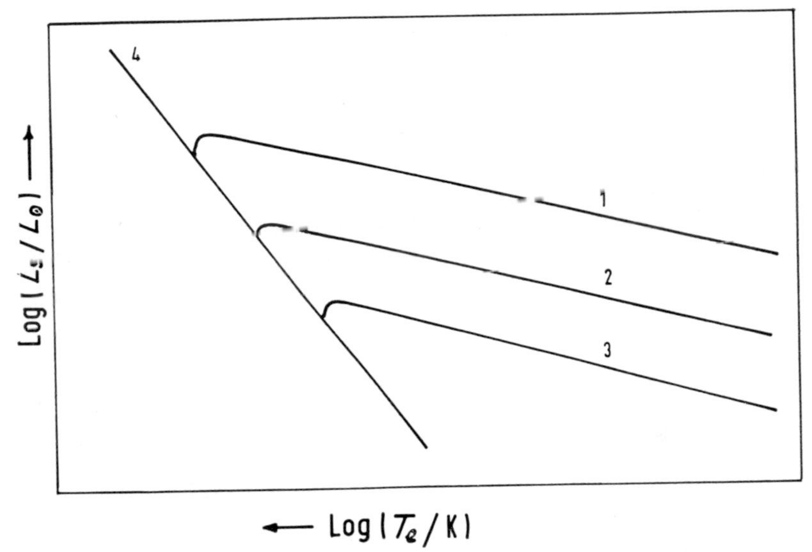

Fig. 51. The approach to the main sequence of fully radiative stars. The curves 1, 2, 3 refer to stars of three different masses and the line 4 is the main sequence.

detail, it was found that they had some defects. In the first instance it was found that they contained important regions in which the criterion

$$\frac{P\,\mathrm{d}T}{T\,\mathrm{d}P} > \frac{\gamma - 1}{\gamma}, \tag{3.65}$$

for convective instability *was* satisfied. The assumption that all of the energy was carried by radiation was incorrect. In addition it was found that use of the boundary condition

$$\rho = 0, \quad T = 0 \quad \text{at} \quad M = M_s \tag{3.77}$$

was inadequate.

Improved surface boundary condition

The visible surface of a star is the level from which radiation can just about escape without any further absorption. The temperature of the visible surface has been assumed to be approximately T_e; that is the meaning of the equation:

$$L_s = \pi acr_s^2 T_e^4. \tag{2.7}$$

A solution of the stellar structure equations obtained using the approximate surface boundary condition (3.77) will be a good solution provided radiation can just about escape from the level where $T = T_e$. If we suppose that in such a solution $T = T_e$ at $r = r_e$ while $T = 0$ at $r = r_s$, radiation will just about escape from the level where $T = T_e$ if

$$\int_{r_e}^{r_s} \kappa\rho \, dr \simeq 1. \tag{5.38}$$

(This should be compared with equation (4.29). A more detailed study gives $\frac{2}{3}$ on the right-hand side of (5.38).) For many stars (5.38) is satisfied by solutions of the equations in which the simple boundary condition (3.77) has been used. However, for protostars, and in fact any stars which have deep outer convection zones, it is necessary to use a more accurate boundary condition based on (5.38).

Hayashi's theory of pre-main-sequence evolution

When Hayashi introduced the better surface boundary condition and allowed for the existence of energy transport by convection, he found

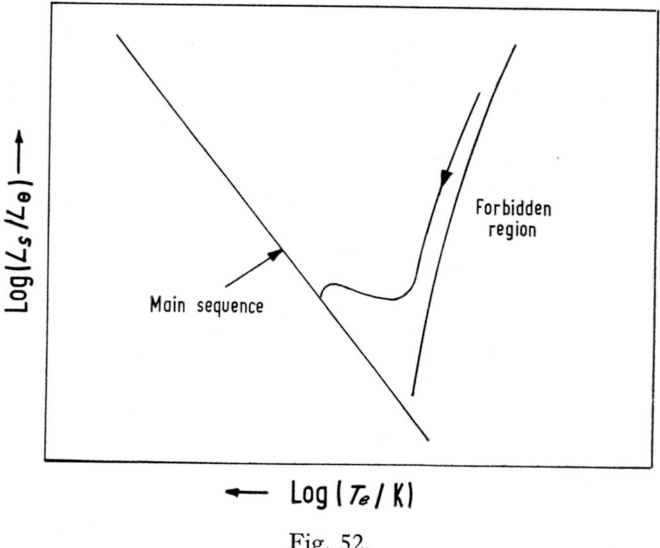

Fig. 52.

that convection could be important throughout the whole of a protostar. More unexpectedly he found that there was a region in the HR diagram in which he could find no reasonable solutions to the equations of stellar structure. This is known as the *Hayashi forbidden zone* and it is shown in fig. 52. Also shown is his result for the immediate pre-main-sequence evolution of a star in which it can be seen that he predicts that stars can have a pre-main-sequence luminosity much higher than their main sequence luminosity. Stars just at the boundary of the Hayashi forbidden zone have convection occurring throughout their interiors and stars to the left of the forbidden region have regions in which convection is not occurring.

If the immediate pre-main-sequence luminosity can be much higher than the main sequence luminosity, it is important to try to understand why this is so and how the immediate pre-main-sequence state can be reached from the initial state of low luminosity and surface temperature. Hayashi's work gives a mathematical answer but a physical understanding is also desirable. In the early stages of the evolution of a protostar its luminosity is determined by how rapidly it can radiate energy and there is no reason why this rate of loss of energy should bear any relation to its main sequence luminosity. In the earliest stages the stellar material is transparent rather than opaque to radiation and the luminosity increases rapidly as the star contracts. Eventually the star becomes opaque and traps radiation within it and at this stage there is a decrease in luminosity. Throughout these initial stages the stellar material has radiated energy so efficiently that its temperature has probably been between 10 K and 20 K.

Once the star is opaque its internal temperature rises and there comes a time when further increase of temperature causes first molecular hydrogen to be dissociated and then atomic hydrogen to be ionized. Both these changes of state require a considerable amount of energy which can only come from the gravitational energy of the star and they trigger off another stage of rapid contraction in which there is another substantial increase in luminosity. Finally when essentially all of the material is ionized the rapid collapse stops and the star enters the final approach to the main sequence shown in fig. 52.

One calculation of the approach to the main sequence for a star of solar mass has been made by Hayashi and his colleagues and it is illustrated in fig. 53†. Although this is a fairly detailed discussion of pre-main-sequence evolution it cannot be regarded as a conclusive study. Other workers have made calculations which give results which are very different from those of Hayashi. Interstellar gas clouds are observed to have rotational energy and to be pervaded by interstellar magnetic fields and both of these properties may seriously affect star formation and pre-main-sequence evolution, but discussion of these factors is outside the scope of the present book. In addition some stars which are believed to be approaching the main sequence are found to be variable stars of the T Tauri type. These are irregular variables which are losing mass to the interstellar medium and such mass loss may play

† It appears that the results of fig. 53 contradict the existence of the Hayashi forbidden zone. However, Hayashi's result was obtained on the assumption that very rapid time variations were not occurring and they do occur in all the early phases shown in fig. 53.

an important role in pre-main-sequence evolution. There is certainly still much work left to be done on the subject of star formation.

There has been a large amount of recent work on the final stages of pre-main-sequence evolution and one set of results due to Iben are shown in fig. 54. It can be seen that the final stage of the approach to

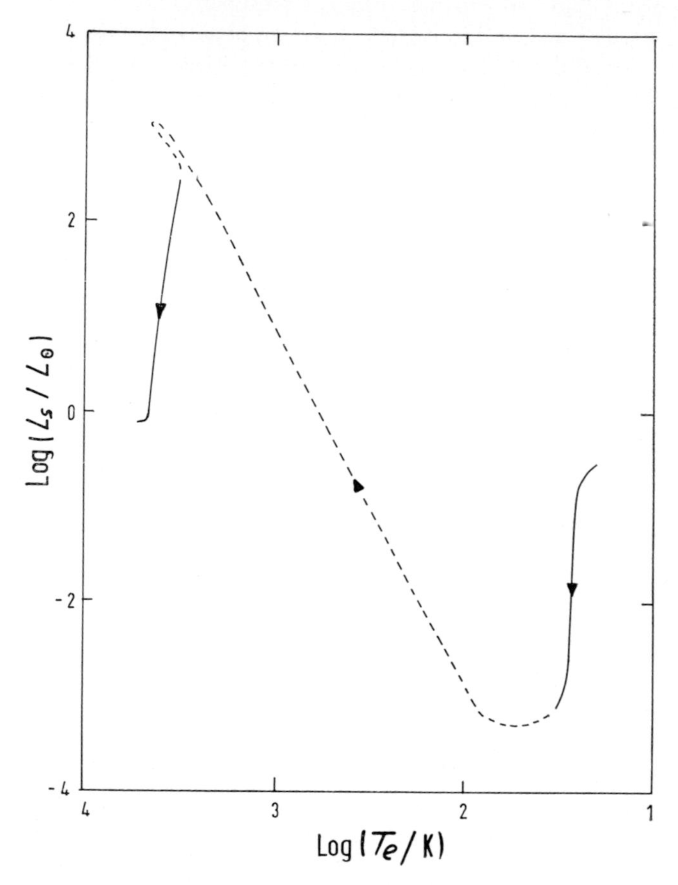

Fig. 53. Pre-main-sequence evolution of a star of solar mass.

the main sequence is similar to the results shown for completely radiative stars in fig. 51, provided only that the main sequence position of the star is well to the left of the Hayashi forbidden region. This is true for massive stars and for them the pre-main-sequence luminosity peak is certainly less significant than for low mass stars.

The whole of pre-main-sequence evolution is rapid compared to the main sequence lifetime. In equation (3.40) we estimated how long the Sun could have radiated at its present rate if its only energy supply had been from gravitational contraction and obtained the result:

$$t \simeq GM_\odot^2/L_\odot r_\odot \simeq 10^{15}\ \text{s} \simeq 3\times 10^7\ \text{years.} \tag{3.40}$$

At that stage we made this estimate to demonstrate that the Sun must have had a different energy source during the past 10^9 years. During the pre-main-sequence phase the main source of energy is gravitational

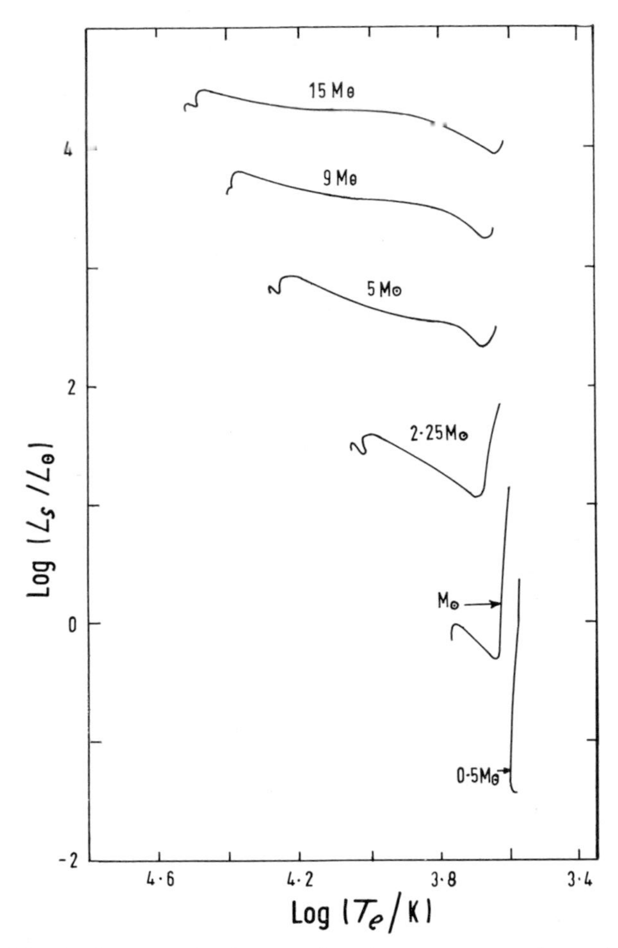

Fig. 54. The final approach of stars of various masses to the main sequence.

contraction so a first estimate of how long it took the Sun to reach the main sequence is given by (3.40). Now we believe that the Sun may have been 500 times as luminous as it is now at some stage in the past, it seems possible that the pre-main-sequence age could be significantly less than the value given in (3.40). In fact, for stars of a solar mass and greater, this reduction does not occur. It has been realized that the initial reactions of the CN cycle (4.16), which convert ^{12}C into ^{14}N, occur before the main sequence is reached and increase the time of approach to the main sequence. The effect of these reactions on the overall chemical composition of a star is slight and this does not affect our assertion that stars reach the main sequence with essentially their original chemical composition.

The calculated times to reach the main sequence for Iben's models are shown in Table 7. For massive stars the time to reach the main sequence is longer than that given by the analogue of equation (3.40) because the additional time allowed by the nuclear reactions between the light elements is greater than the reduction due to the high luminosity phase. For the star of one solar mass, Hayashi's calculations show that the entire evolution before the maximum luminosity is reached lasts

Mass	15·0	9·0	5·0	3·0	2·25
Time	$6{\cdot}2\times10^4$	$1{\cdot}5\times10^5$	$5{\cdot}8\times10^5$	$2{\cdot}5\times10^6$	$5{\cdot}9\times10^6$
Mass	1·5	1·25	1·0	0·5	
Time	$1{\cdot}8\times10^7$	$2{\cdot}9\times10^7$	$5{\cdot}0\times10^7$	$1{\cdot}5\times10^8$	

Table 7. Time taken to reach main sequence for stars of different masses (mass in $M_\odot$ and time in years).

only about 20 years. Because the pre-main-sequence evolution is very rapid compared to the main sequence lifetime, we are likely to observe only a relatively small number of stars in the stage of pre-main-sequence contraction if stars have been forming steadily during the lifetime of the Galaxy. Thus, in this case, *the number of stars observed in a given phase of evolution should be roughly proportional to the time an individual star spends in that phase.* The simplest example of this is that most stars are main sequence stars because stars spend a large portion of their life on the main sequence. This will be discussed in the context of post-main-sequence evolution in Chapter 6. It seems very unlikely indeed that we shall be able to identify any stars as being in the stage of evolution before the final luminosity maximum, although one such identification has been tentatively suggested. This is one of the basic difficulties with trying to observe stellar evolution. *Normally evolution occurs so very slowly that observation of it is quite impossible. When evolution is rapid the phase of evolution is soon over and then the statistics are against our finding any stars in that phase.* One exception, where rapid evolution is observed, is the explosion of a supernova. These stars become so highly luminous that they draw attention to themselves even at very large distances.

One consequence of our present views on the pre-main-sequence evolution of the Sun should be mentioned. It is thought that the Earth and the other planets were formed out of solar material or at least out of the same material as that from which the Sun was formed. Until Hayashi's work it was always assumed that during the period of planetary formation the Sun was radiating less energy than it is today. Now it seems likely that the Sun was much more luminous for part of the time and this might have important effects on theories of the origin of the solar system.

Summary of Chapter 5

In this chapter we have investigated the hypothesis that main sequence stars are stars of uniform chemical composition which are converting hydrogen into helium in their interiors. Using simple approximations to the laws of opacity and energy generation, which were discussed in Chapter 4, we have shown that such stars obey a mass–luminosity relation similar to that of main sequence stars and that they lie in a region in the HR diagram in qualitative agreement with the observed main sequence. Results of more detailed calculations, with more accurate mathematical expressions for the laws of opacity and energy generation, confirm these qualitative results. Red giants and white dwarfs are not stars of uniform chemical composition which are radiating energy which has been released in nuclear reactions.

Main sequence stellar structure is essentially independent of the star's previous life history and this is fortunate as the theory of star formation and pre-main-sequence stellar evolution still contains many uncertainties. Stars must initially be large, cool and of low luminosity and in approaching the main sequence they must become smaller, hotter and brighter. It is now believed that the luminosity of a star may have a maximum value in the pre-main-sequence phase which is much greater than its main sequence value.

CHAPTER 6

early post-main-sequence evolution and the ages of star clusters

Historical introduction

AFTER the main sequence the most prominent group of stars in the HR diagram (fig. 55) is the red giants and supergiants. These stars have larger luminosities and radii than main sequence stars of the same colour. From the discussion in Chapter 5, it appears that red giants are not stars of homogeneous chemical composition and we must now discover

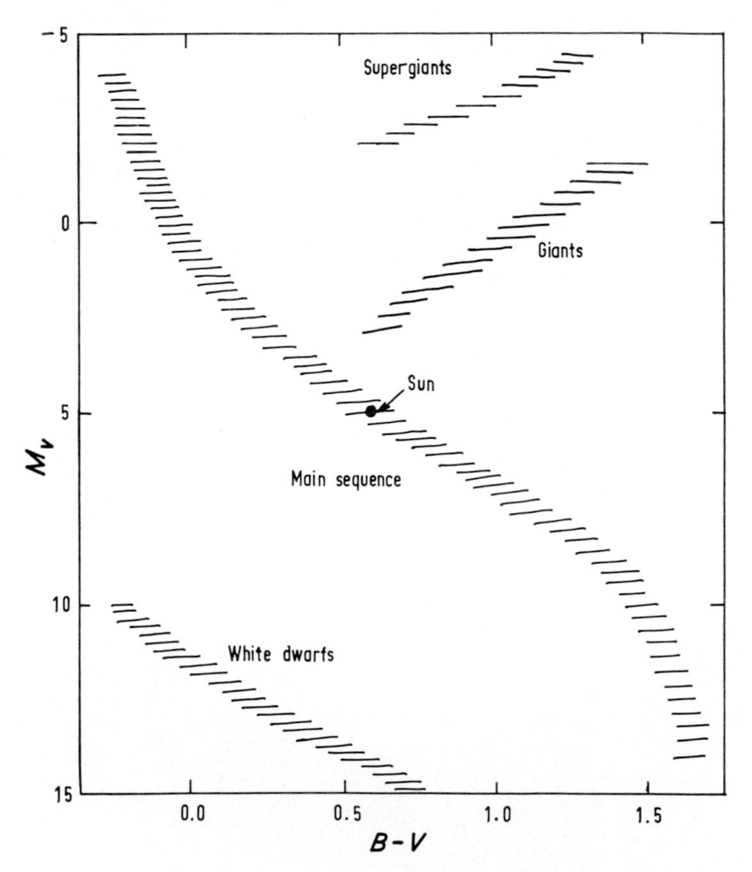

Fig. 55. The Hertzsprung–Russell diagram for nearby stars.

how red giants differ from main sequence stars in their internal structure as well as in their surface properties. We have already indicated at the end of Chapter 2 (fig. 30) that stars become red giants when nuclear reactions in their interiors lead to a non-uniformity of chemical composition. Before we discuss this further, we will give a brief historical introduction to the problem of the red giants. Although in this book we mainly discuss the present state of knowledge, it is perhaps instructive in one case to trace the steps by which the present knowledge has been obtained.

When the first theoretical calculations of stellar structure were made, it was very difficult to explain the occurrence of red giants, since at that time it was believed that stars remained chemically homogeneous as they evolved. As will now be described, it was believed that the rotation of

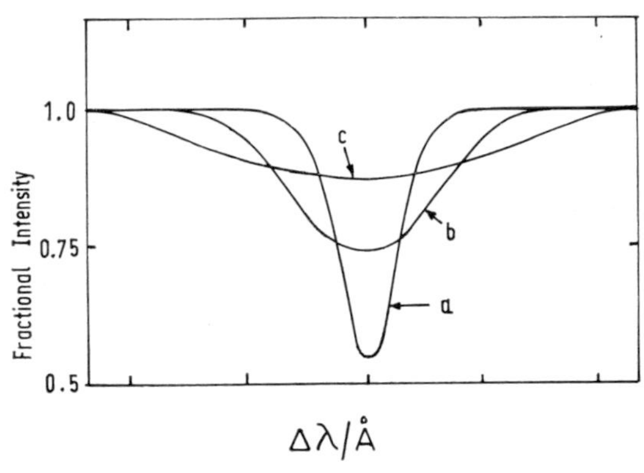

Fig. 56. The effect of rotation on the shape of a spectral line. a is the shape of the spectral line in a non-rotating star, b, in a star of moderate rotation speed and c, in a rapidly rotating star.

stars caused them to be well mixed. Most stars are observed to rotate, even if the rotation of many of them is not sufficiently rapid to distort their structure substantially. Rotation is detected by the Doppler effect. As has previously been mentioned, radiation from a source moving away from an observer is shifted to the red while that from a source moving towards the observer has a blue shift. If a star is rotating, part of it is moving towards us and part away from us and this causes any spectral line from the star to be broadened (fig. 56). The observation of broadened spectral lines leads to the deduction that stars are rotating.

A rotating star is not spherical and the surfaces of constant temperature, density and pressure in such a star are spheroids to a first approximation (fig. 57). In this figure it can be seen that the temperature gradient near the poles of such a star is greater than the temperature gradient near the equator as the surfaces of constant temperature are closer at the pole. This in turn means that more energy is carried by radiation near the pole and left to itself this would disturb the surfaces

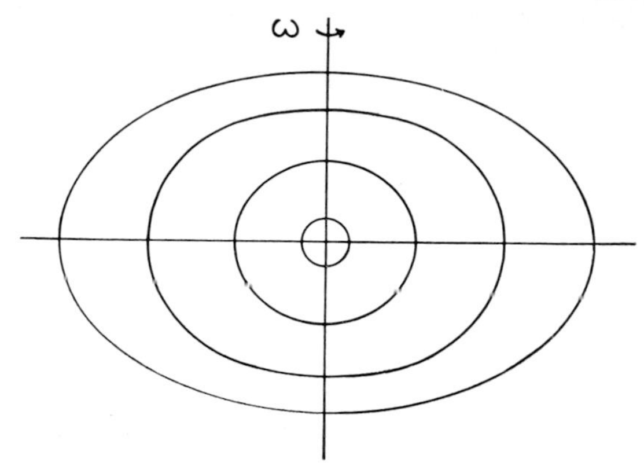

Fig. 57. Rotating star. The surfaces shown have constant pressure and temperature.

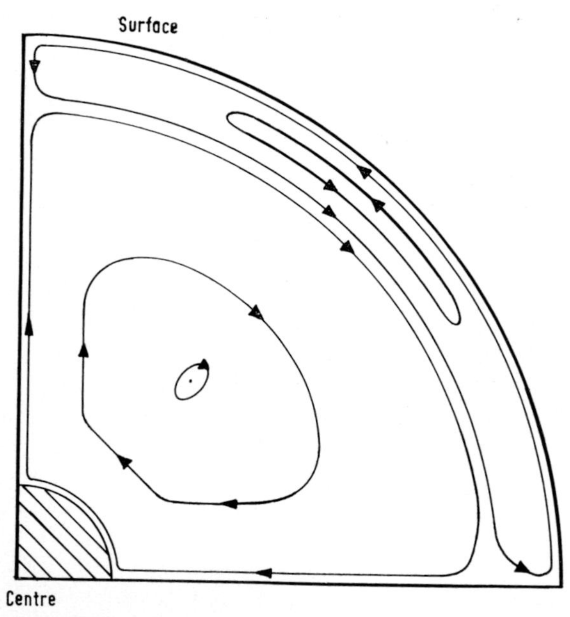

Fig. 58. Meridional circulation currents in a rotating star. The hatched area is a convective core and the arrows show the direction of the currents.

of constant temperature. Eddington showed that the surfaces of constant temperature are preserved by slow circulating motions of the type shown in fig. 58 which carry both material and energy from one part of a star to another. These are known as *meridional currents* and Eddington believed that they kept a star chemically homogeneous as it evolved. The typical speed of the currents is now believed to be:

$$v \approx (\omega r_s^2/g)(L_s/M_s g), \tag{6.1}$$

where ω is the angular velocity of the star and g the acceleration due to gravity, but Eddington's original estimate was several orders of magnitude greater than this. As a result, Eddington believed that even slowly rotating stars would be well mixed. Although the meridional circulation produces some of the same effects as convection it is qualitatively rather different; it is produced by rotation rather than by a large temperature gradient and the motions are regular on a very large scale, while convective motions are very irregular. In addition, in any

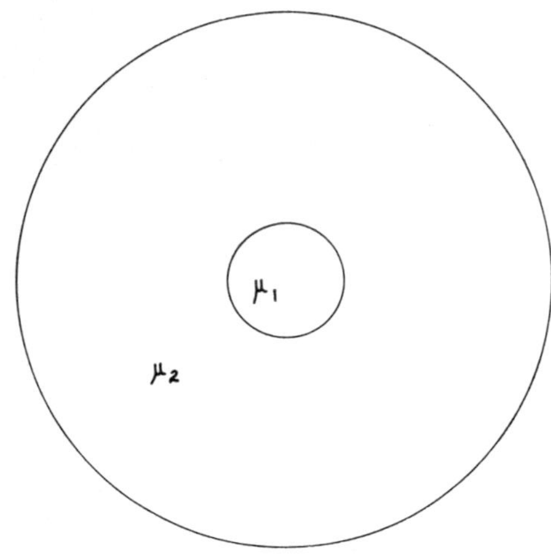

Fig. 59. Star with discontinuous chemical composition. A region with molecular weight μ_1 is surrounded by a region with molecular weight μ_2.

region in which convection is carrying energy, the convective motions are very much more rapid than the meridional circulation.

If hydrogen burning stars remained chemically homogeneous as they evolved they would remain in the neighbourhood of the main sequence. As we have seen in Chapter 5 (fig. 48) the conversion of hydrogen into helium is accompanied by a motion to the left and upwards in the HR diagram and not into the region occupied by the red giants. As red giants did not appear to occur naturally in the process of stellar evolution, theoretical astrophysicists experimented with models which might have *giant* properties. They found that models with a single discontinuity of chemical composition (fig. 59), with the inside region of higher molecular weight, could have large radii provided that the ratio of masses in the two zones was chosen carefully. It was then necessary to think of ways in which this discontinuity of chemical composition could be produced.

F. Hoyle and R. A. Lyttleton proposed that a star passing through an interstellar gas cloud could increase its mass by *accretion* of some of the material of the cloud. If the star had already converted some of its hydrogen into helium and if the interstellar gas cloud consisted mainly of hydrogen, a discontinuity of chemical composition would be produced and the star might become a red giant. It would then remain a red giant until mixing currents once again made the star homogeneous. According to this picture a star would spend a limited time as a red giant and it might become a red giant several times in its life history.

At the time that this suggestion was made, the clouds of neutral (un-ionized) hydrogen in the Galaxy had not yet been discovered. They were subsequently detected by radio astronomers because even very cold hydrogen emits radio waves at a wavelength of 21 cm (0·21 m). When the distribution of neutral hydrogen clouds in the Galaxy was mapped by the radio astronomers, it was found that these clouds were neither dense enough nor slowly moving enough† to permit substantial accretion of interstellar matter by stars. In addition, it was difficult to see how the rather well-defined giant branches of galactic and globular clusters could be produced if matter was being accreted by stars of all masses in the clusters. It was fortunate that, just at the time that accretion theory became untenable, an error was discovered in the original estimates of the speed of meridional circulation currents made by Eddington. With the revised speeds, it appears that most stars will not be mixed by meridional circulation. For example, in the interior of the Sun the circulation currents are believed to travel at 10^{-11} m s^{-1} and at this speed they take more than 10^{12} years for one circulation, while substantial changes of chemical composition, due to nuclear reactions, can occur in the solar interior in less than 10^{10} years. It thus appears that natural inhomogeneities of chemical composition can arise as a star evolves and can lead to its becoming a giant. This concludes this historical introduction and we now describe our present state of knowledge.

General character of post-main-sequence evolution

We shall find that the details of post-main-sequence evolution depend on stellar mass and, in particular, on whether or not a star has a convective core when it is on the main sequence. If a main sequence star has a convective core, material from throughout the convective core can be carried into the central regions and is available for nuclear reactions, regardless of the temperature difference which exists between the centre

† The cloud densities can be deduced from a measurement of the total 21 cm emission from a given volume while the cloud velocities are found from the Doppler effect on the 21 cm line.

and the surface of the core. If the star has a core in which all of the energy transport is by radiation, there will be no mixing processes in the interior of the star. Thus the exhaustion of the central supply of hydrogen will be determined entirely by the speed of nuclear reactions at the centre and not by the rate at which unburnt hydrogen can be carried into the centre.

Broadly speaking the first crucial stage in the evolution of stars away from the main sequence occurs when the central hydrogen content falls to zero. Before this stage, calculations show that the star's luminosity gradually increases but that it remains in the general neighbourhood of the main sequence. Once there is no hydrogen left in the centre, the

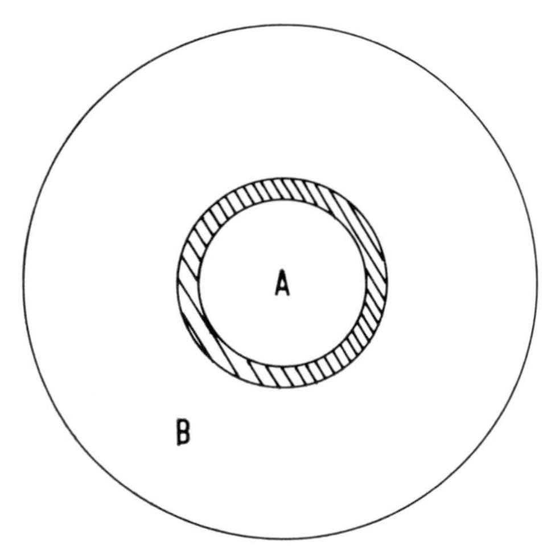

Fig. 60. Star with hydrogen burning shell. In region A all of the hydrogen has been converted into helium, in region B no nuclear reactions have occurred and in the hatched area hydrogen is burning into helium.

central regions stop releasing nuclear energy. Instead they resume their release of gravitational energy and begin to contract, slowly in the first instance. As this happens, the region in which hydrogen is being burnt gradually moves outward through the star and what is known as a hydrogen burning shell is produced (fig. 60). As the only energy release in the central regions of the star is now gravitational potential energy, the luminosity in that region is very low. Because the luminosity is low, only a very small temperature gradient is needed to carry the energy outwards (see equation (3.51) for example) and the star has an almost *isothermal core* composed of helium and a small admixture of heavy elements, corresponding to the initial heavy element content of the star.

As the evolution of the star is followed, it is found that a dramatic change occurs when the isothermal core contains between 10% and

15% of the mass of the whole star. It is found that the pressure gradient in a larger isothermal core, which is contracting slowly, is unable to support the outer regions of the star and the central regions now collapse rapidly instead of contracting slowly. This critical mass for a slowly contracting isothermal core is known as the *Schönberg–Chandrasekhar limit* and we shall see later in this chapter that it is very important in the understanding of the galactic cluster HR diagrams.

When calculations are continued past the stage at which the Schönberg–Chandrasekhar limiting mass is reached, it is found that the inner layers of the star contract rapidly and heat up, but that simultaneously the radius of the star as a whole expands. The heating of the inside is produced by the rapid release of gravitational energy. Although the surface expansion which accompanies the core contraction is predicted by the solutions of the equations of stellar structure for a star with variable chemical composition and although such an expansion is

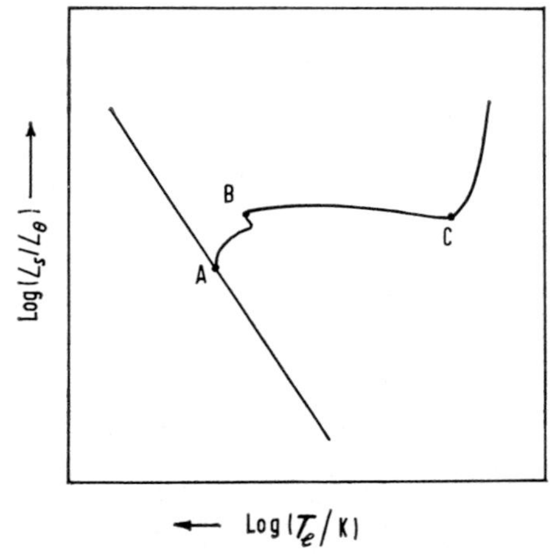

Fig. 61. Evolution to the giant region. An isothermal core is formed near B and a deep outer convection zone appears near C.

necessary to explain the existence of red giants, it is not easy to give a simple explanation of why it occurs. There have been attempts at plausible explanations such as the contraction of the hydrogen burning shell leading to a very concentrated release of energy which can only escape from the star if the overlying layers are expanded to reduce the effective opacity. However, none of the simple explanations is entirely convincing, but there is in any case no reason to doubt the properties of the solutions of the equations of stellar structure which do predict that stars should expand and become red giants. The calculations predict that the star expands without any significant change in luminosity and this means that the star moves rapidly to the right in the HR diagram (fig. 61).

Is there anything which will stop this contraction of the core and expansion of the surface layers? There are, at least, three ways in which the movement to the right in the HR diagram can be halted or slowed down. In the first case, if the central temperature of the star continues to rise, as it will according to the Virial Theorem if the central material remains a perfect gas, it may become high enough for the next nuclear fuel, helium, to be burnt through the reaction discussed in Chapter 4 (equation (4.27)). When this happens, the isothermal core is replaced by a helium burning convective core and, with the renewed release of nuclear energy, the release of gravitational energy and the central collapse are stopped. In the second case the central regions may become so dense that the pressure in the central regions is given by the degenerate gas law rather than the perfect gas law. At this stage further contraction becomes more difficult because of the high dependence on density of the pressure of a degenerate gas:

$$P_{\text{gas}} \simeq K_1 \rho^{5/3}. \tag{4.49}$$

Finally, if the surface temperature of the star becomes very low, there is once again a deep outer convection zone because of the existence of ionization zones of hydrogen and helium just below the stellar surface. At about this stage the movement of the star to the right in the HR diagram is halted. This is because any further movement to the right would take the star into the Hayashi forbidden zone which was discussed in Chapter 5. The star does not cease to expand, but further increase in radius is now accompanied by an increase in luminosity rather than by a decrease in surface temperature ($L_s = \pi acr_s{}^2 T_e{}^4$). This is also shown in fig. 61.

Dependence of early evolution on stellar mass

Evolution of stars of high mass. As mentioned earlier, the exact way in which a star evolves depends on its mass, and we now give a general description of the mass dependence. High mass stars have large convective cores when they are on the main sequence; the size of the convective core as a function of stellar mass has already been shown in Table 6 of Chapter 5. For such high mass stars, the supply of hydrogen which can be used in an initial burning phase is very large, because convection currents keep the entire convective core mixed. For very massive stars ($\gtrsim 10 M_\odot$) the mass in the convective region may grow as the star evolves, but there is some dispute about this at present. Even if the mass in the core decreases as the star evolves, a considerable fraction of the star's original hydrogen content may be burnt before the star exhausts its central hydrogen. When this does happen, the newly formed isothermal core may be almost immediately as large as the Schönberg–Chandrasekhar limit, so that the core collapses rapidly without an initial period of slow contraction.

Figure 62 shows how the hydrogen in the central regions of such a star is depleted as the star evolves. The hydrogen content X is plotted against fractional mass ($m \equiv M/M_s$) for several stages in the star's evolution. At each stage there is a convective core of uniform chemical composition, surrounded by an intermediate zone of variable chemical composition and an outer region with the original chemical composition of the star. The intermediate zone contains material which was in the convective core when the star was on the main sequence and its hydrogen content has been reduced by convection currents mixing it with material in which nuclear reactions have occurred.

During these phases of evolution the star does not move very far from the initial main sequence in the HR diagram, although it does show a

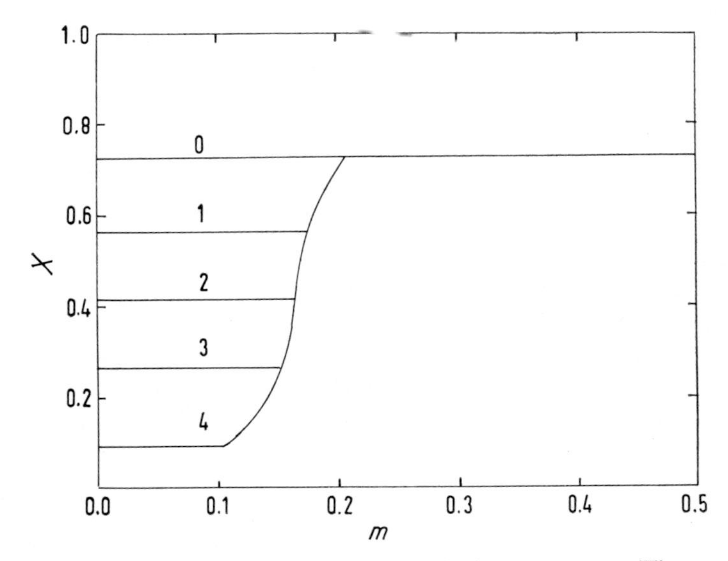

Fig. 62. The depletion of hydrogen in a high mass star. The numbers denote successive stages in the star's evolution.

slight increase in luminosity and decrease in surface temperature and this process leads to the appearance of a finite width in the observed main sequence. Thus from the surface properties of a star alone it is impossible to tell whether the star has just reached the main sequence or whether it has almost used up its central hydrogen and has moved some way from its initial position on the main sequence. This is one reason why the main sequence of the nearby stars has a finite width but it is not the only one. As we have seen in Chapter 5, the position of the main sequence varies with the chemical composition of the stars and as all stars do not have the same chemical composition this gives a finite width to the observed sequence. In addition, a study of the structure of rapidly rotating stars, which is outside the scope of this book, shows that they have somewhat different luminosities and effective temperatures from non-rotating (or slowly rotating) stars of the same mass and chemical composition.

Once hydrogen is exhausted in the central regions, a nearly isothermal core of mass greater than the Schönberg–Chandrasekhar limit is soon produced and the star moves rapidly to the right in the HR diagram. What observational consequences can we expect from this rapid evolution? As we have already mentioned in Chapter 5, if stars travel through a particular region in the HR diagram in a very short time, we cannot expect to see many stars in that region at any given time. This immediately gives us a qualitative explanation of one of the properties of galactic cluster HR diagrams which we have already discussed in Chapter 2 (fig. 26) page 39. In fig. 26 it can be seen that in those clusters which possess main sequence stars of high luminosity, and hence presumably high mass, there is a gap called the Hertzsprung gap between the main sequence and the red giants. Stars presumably cross that region very rapidly immediately after central hydrogen exhaustion and that is why very few stars (if any) are found in the Hertzsprung gap. The HR diagram for the nearby stars (fig. 22) shares the same property in that at high luminosity there is a distinct gap between the main sequence and the giant branch.

Evolution of low mass stars

We next consider low mass stars. These have only a small convective core or no convective core at all. This means that, when the hydrogen has been used up in the stellar core, it is only exhausted in a very small central region (fig. 63). As a result a small isothermal core is formed and it starts to contract long before its mass is comparable with the Schönberg–Chandrasekhar limit. It follows that the movement away from the main sequence is less rapid for low mass stars than for high mass stars and in addition the onset of degeneracy and a deep outer convection zone occur sooner for low mass stars, since they have higher central densities and lower surface temperatures than high mass stars before the central contraction and the outer expansion starts. In figs. 25 and 26 of Chapter 2, we have seen that there is no Hertzsprung gap in the HR diagrams of both globular clusters and old galactic clusters in which stars leaving the main sequence have low luminosity and hence presumably low mass. This is explained by the relatively slow evolution away from the main sequence for low mass stars. This will be discussed further below, when the significance of the phrase *old* galactic cluster will also become clearer.

The end of early evolution

Before presenting detailed results of recent calculations, it is perhaps necessary to define what we mean by early post-main-sequence evolution. The definition we adopt is not necessarily one of astronomical significance. We shall use the phrase early evolution to mean evolution from the main sequence until we reach a stage in which there is a real uncertainty in our theoretical calculations. We might, for example, reach a stage where, either because of lack of knowledge of the laws of physics or because of the inadequate size or speed of present computers, it is impossible to follow the evolution of the star any further.

One such specific uncertainty arises when the central regions of a star become degenerate, but not sufficiently degenerate to prevent the central temperature from continuing to rise to a value when the next set of energy releasing nuclear reactions occur. The onset of nuclear reactions

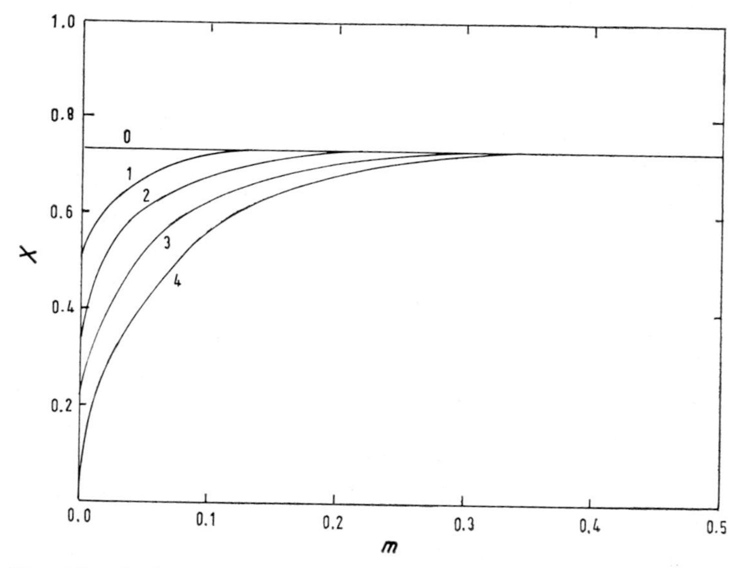

Fig. 63. The depletion of hydrogen in a low mass star. The numbered curves refer to successive stages in the star's evolution. Although hydrogen is initially exhausted only in a very small central region, some hydrogen burning occurs further out than in a star of higher mass. This is caused by the relatively small temperature dependence of the rate of energy generation by the PP chain.

in degenerate material can be very different from the onset of nuclear reactions in non-degenerate material. If nuclear burning starts in a perfect gas, this is potentially stable in the following sense. Suppose a small temperature rise occurs in such a region causing a large release of additional energy, because the rate of nuclear reactions depends on a high power of the temperature. If the stellar material is fairly opaque, this energy may not be able to escape as rapidly as it is produced. In that case a further local temperature rise will occur, but in a perfect gas this will also cause the pressure to increase and this will, in turn, lead to an expansion and cooling and a reduction in the rate of energy release. As an example of this process, we have already seen in Chapter

5 that, when stars approach the main sequence, the initial luminosity when nuclear reactions start is slightly higher than the luminosity when the star is on the main sequence.

If nuclear burning starts in a degenerate gas, the consequences can be quite different. The argument given above is still true as far as the further temperature rise. However, as we have discussed in Chapter 4, the pressure of a degenerate gas scarcely depends on temperature and this rise in temperature leads to a negligible pressure increase which is quite inadequate to cause expansion and cooling. In this case the local temperature rise, with a rapid increase in the release of energy and hence the rate of rise of temperature, continues until the temperature is high enough for the material to become non-degenerate. At this stage the material will expand, but, because of the runaway release of nuclear energy, the expansion is likely to be explosive. In some cases the observed explosions of stars may be due to the ignition of a nuclear fuel in a degenerate gas. It is not, however, clear that an explosion of the central regions of a star will necessarily lead to an explosion of the visible regions. For this to happen the central explosion must be intense enough to set the whole of the overlying layers of the star into violent motion.

In low mass stars, helium burning starts in material which is already degenerate and the ignition of helium in such circumstances is called the *helium flash*. The very high temperature dependence of the energy release from helium burning

$$\varepsilon_{3\mathrm{He}} = \varepsilon_3 X_{\mathrm{He}}{}^3 \rho^2 T^{40}, \tag{4.28}$$

makes an explosive release of energy very likely. When such a helium flash occurs, it is very difficult to solve the equations of stellar structure accurately and no completely satisfactory study of the evolution of low mass stars past the onset of helium burning has yet been completed. Whereas in most stages of evolution nothing significant will happen to the structure of a star of one solar mass in a period of a million years, when the helium flash starts significant changes occur in the central regions in 100 s. Even with a very large computer it is difficult to study the changes in the structure of a star in such rapid evolutionary phases. For such low mass stars, the onset of helium burning can be taken to mark the end of early post-main-sequence evolution. In more massive stars it appears that helium burning starts uneventfully and it is possible to study the evolution of the star until all of the central helium has been turned into carbon. If the central regions then become degenerate before the temperature is high enough for nuclear reactions burning carbon to occur, it is possible that an explosive energy release may occur at the onset of carbon burning.

Detailed calculations of post-main-sequence evolution

Relatively massive stars ($3M_\odot \leqslant M \leqslant 10M_\odot$). Recent calculations of the evolution of relatively massive stars by I. Iben are illustrated in fig. 64. In these calculations an initial chemical composition has been assumed for all of the stars which is similar to the chemical composition of population I stars in the Galaxy. In detail it is:

$$X = 0{\cdot}708, \quad Y = 0{\cdot}272, \quad Z = 0{\cdot}020. \qquad (6.2)$$

The reason behind this choice is that any relatively massive star which is in the phase of early post-main-sequence evolution must have been formed quite recently in the galactic lifetime because its main sequence lifetime is quite short (see Table 5). It appears that population I stars are relatively young stars and that they have a higher content of heavy

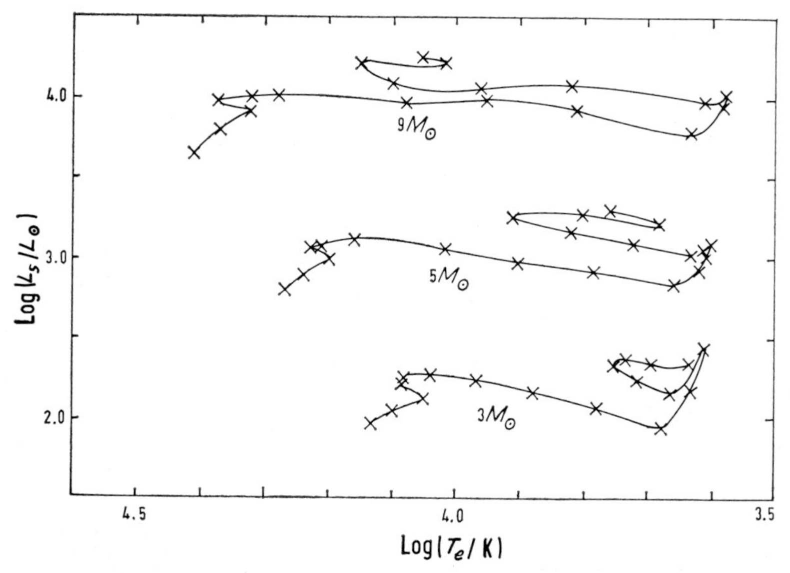

Fig. 64. Post-main-sequence evolution of relatively massive stars.

elements than the older population II stars. In the calculations it has been assumed that the stars evolve with constant mass and the best available values have been used for laws of opacity and energy generation. It can be seen from fig. 64 that the evolutionary track of a single star is very complicated according to present theories.

R. Kippenhahn and his collaborators have also studied the evolution of relatively massive stars and they have followed them to a later stage in their evolution. They have used a different chemical composition from that chosen by Iben:

$$X = 0{\cdot}602, \quad Y = 0{\cdot}354, \quad Z = 0{\cdot}044. \qquad (6.3)$$

The results obtained by Kippenhahn are shown in fig. 65. Where they can be compared with the results obtained by Iben, they are generally similar although there are some detailed differences. A comparison of the curves in figs. 64 and 65 gives an idea of what we believe to be the uncertainties in the theory of the evolution of stars in this mass range. Kippenhahn's group has also considered the evolution of stars with different chemical compositions and they have found that

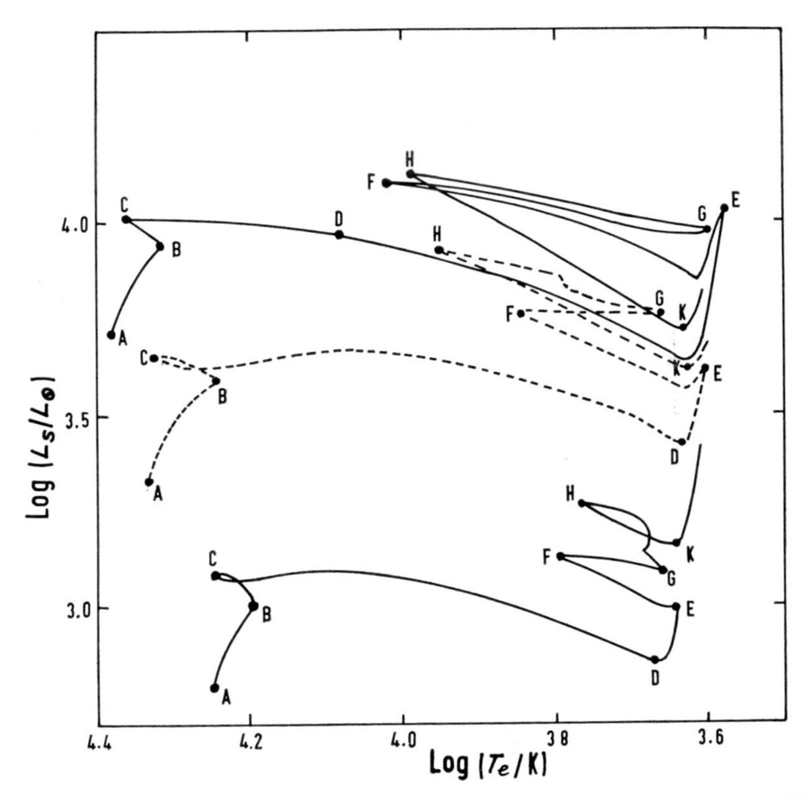

Fig. 65. Post-main-sequence evolution of relatively massive stars. The stars have a different chemical composition from those of fig. 64 and are followed through a greater fraction of their life history. From the top the curves are for $9M_\odot$, $7M_\odot$ and $5M_\odot$.

quite small changes in the value of Z can lead to substantial changes in the evolutionary tracks. Their results for a different chemical composition

$$X = 0{\cdot}739, \quad Y = 0{\cdot}240, \quad Z = 0{\cdot}021 \tag{6.4}$$

are shown in fig. 66. It can be seen that there are some substantial differences between these results and those of Iben.

Although the individual evolutionary tracks of stars are very complicated, according to current theories, this does not mean that these complications will all be reflected in the HR diagram of a star cluster.

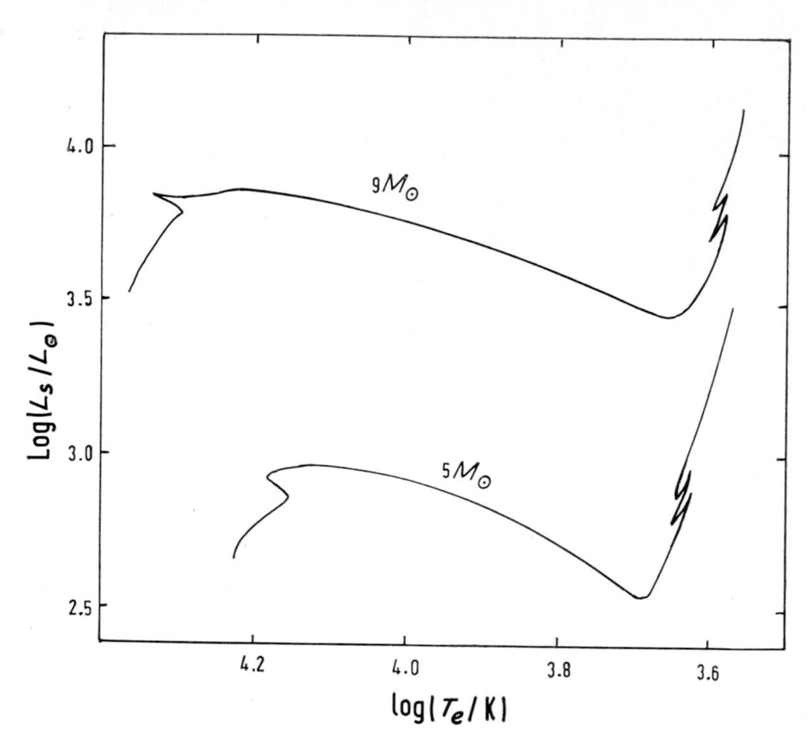

Fig. 66. Post-main-sequence evolution of relatively massive stars, again with a different chemical composition.

$M_s/M_\odot$	B	C	D	E	F	G	H	K
5	5·37	5·62	5·91	5·94	6·76	7·04	7·83	7·86
7	2·56	2·60	2·65	2·66	3·15	3·31	3·56	3·57
9	1·59	1·65	1·66	1·67	1·86	1·94	1·96	1·96

Table 8. Time (in 10^7 years) taken to reach lettered points on evolutionary tracks of fig. 65.

As has been mentioned earlier, some of the evolutionary phases are extremely rapid and this means that there are regions on the evolutionary tracks in which we are very unlikely to see stars. The times taken for different portions of the tracks of fig. 65 are shown in Table 8.

Evolution of a star of five solar masses

For one of the cases studied by Kippenhahn and his colleagues we will discuss the results in much greater detail. This is the star of five solar masses† with chemical composition given in equations (6.3). The evolutionary track of this star is shown in fig. 67 and the condition of the interior of the star at different stages in the evolution is shown in fig. 68. As the evolutionary track of the star is very complicated it is useful also to describe in some detail the state of the star at all of the points labelled in fig. 67. These are:

A: The star is on the main sequence. It has a convective core containing about 21% of the mass of the star and nuclear reactions converting hydrogen into helium are essentially confined to about the inner 7% of its mass.

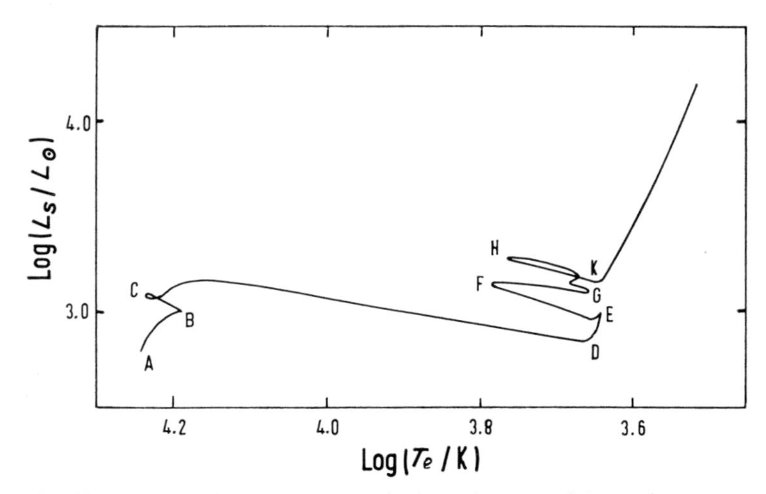

Fig. 67. Post-main-sequence evolution of a star of five solar masses.

B: The convective core has now shrunk to about half of its original size (in terms of mass contained) and a considerable fraction of the central hydrogen has now been consumed.

C: The point of central hydrogen exhaustion. An isothermal core containing no hydrogen is formed and it very soon has a mass exceeding the Schönberg–Chandrasekhar limit, after which rapid collapse of the core occurs. A hydrogen burning region exists outside the isothermal core. This is initially quite thick but, subsequently, as it burns outward in the star, it becomes much narrower.

From C to D the star moves very rapidly to the right in the HR diagram as the isothermal core, more massive than the Schönberg–Chandrasekhar

† It would be nice to refer this discussion to a particular star. However, as mentioned in Chapter 2, there is no giant star with a well-known mass. The two (main-sequence) components of the eclipsing binary U Ophiuci are observed to have masses of about $5M_\odot$ and this discussion can be regarded as a prediction of their future life history.

limit, collapses rapidly and the outer regions expand. The central temperature is too low for helium to burn.

D: At this point the star develops a deep outer convective region which at its maximum extent contains about 54% of the total mass of the star. The star has a structure appropriate to a largely convective star and its movement in the HR diagram becomes almost vertical just to the left of the Hayashi forbidden region; the luminosity and effective temperature are now following a track very similar to the one

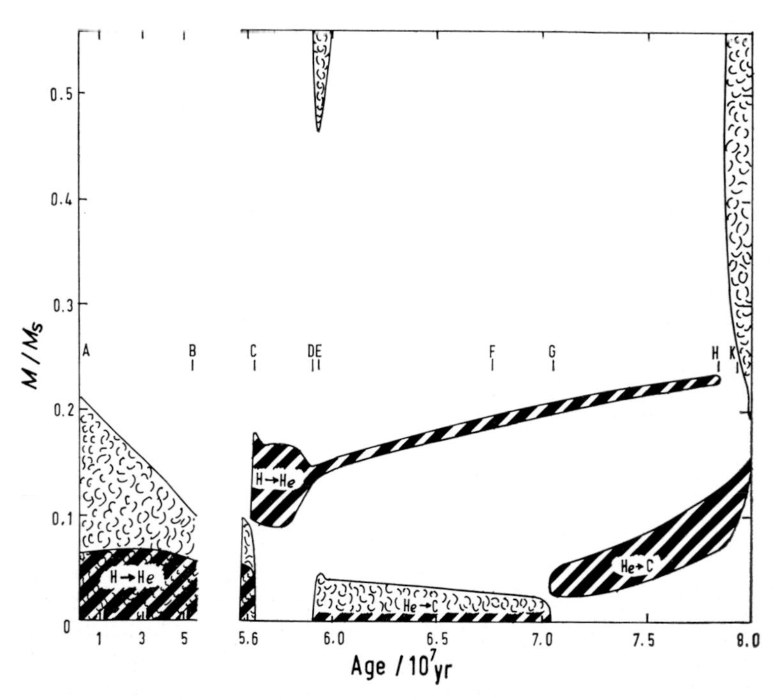

Fig. 68. The internal structure of a star of five solar masses as a function of age. The heavily hatched areas are ones in which the named nuclear reactions are occurring. Convection zones are shown with a *cloudy* appearance.

followed in the opposite direction as the star approached the main sequence (fig. 54), but the internal structure of the star is very different. The hydrogen shell source has become very thin and from here until point H it contains only about 1% of the mass of the star.

E: The point of central helium ignition. Once the helium is burning, the star develops a new convective core containing about 5% of the mass of the star. Helium burns to carbon in the inner part of this convective core. Because the rate of helium burning depends on a very high power of the temperature (see equation (4.28)), significant energy release occurs only in about the inner 1% of the star.

F: By this stage the convective core has shrunk somewhat and the central helium content has been considerably reduced. It is not immediately clear why the star reverses its direction in the HR diagram, although a reversal has already been found at B at a corresponding stage in core hydrogen burning.

G: At this point the central helium content is reduced to zero and the star then has an almost isothermal core composed of carbon and the original admixture of heavier elements, which have not been affected by any of the nuclear reactions. From this stage onwards, helium burns to carbon in a shell outside the isothermal core.

H: Ever since stage C, hydrogen has been burning in a shell which has been gradually moving further out in the star, as can be seen in fig. 68. The temperature of this hydrogen burning shell is determined partly by the properties of that part of the star inside it. At point H this temperature, which has been falling for some time becomes too low for further significant burning of hydrogen into helium and the hydrogen shell source ceases to exist. It may be noted from fig. 67 that, when the shell source does cease to be effective, the star abruptly changes its direction in the HR diagram.

K: The star once again develops a deep outer convection zone, which this time contains more than 80% of the mass of the star. As the star becomes largely convective it once again moves more or less vertically in the HR diagram just to the left of the Hayashi forbidden zone. This convection zone penetrates into the region where all of the hydrogen has been converted into helium and mixes this material with outer layers which still contain their hydrogen. Thus the convection brings hydrogen from the outer part of the star down towards the region where helium is burning. If this hydrogen were brought to too high a temperature, it would ignite explosively. At the time of this work it was uncertain whether or not this would occur before:

L: The ignition in the centre of the star of carbon, which burns through reactions such as

$$^{12}C + {}^{12}C \rightarrow {}^{24}Mg + \gamma, \tag{6.5}$$

$$^{12}C + {}^{12}C \rightarrow {}^{23}Na + p \tag{6.6}$$

and

$$^{12}C + {}^{12}C \rightarrow {}^{20}Ne + {}^{4}He. \tag{6.7}$$

From later calculations it seems that carbon burning probably starts before hydrogen is re-ignited and this presumably leads to the formation of a carbon burning convective core. Although at this stage the material in the centre of the star has become a somewhat imperfect gas, it does not appear that the carbon ignition is explosive.

These results have been described in some detail in order to show what stage studies of stellar evolution have now reached. The evolutionary track of a star appears to be very complicated with multiple passages across the HR diagram. From this and other calculations it appears that each change of direction in the diagram is associated with a change in the importance of some energy source. It should be stressed once again that the time taken in the advanced evolutionary phases is very small compared to the duration of the early phases, so that we do not expect to observe many stars in an advanced stage of evolution and we do not expect all of the complications of an individual evolutionary track to show up in a star cluster HR diagram.

The ages of young star clusters

We now turn to the comparison of theory and observation. We have already mentioned in Chapter 3 that we do not know enough about the properties of individual stars to make it worth while to try to discuss their properties in detail, but we can hope to explain why the HR diagrams of galactic and globular clusters have the shapes which they are observed to have. In particular, we can hope to determine the approximate ages of star clusters, that is how long it is since the stars in the clusters were formed. The shapes of the evolutionary tracks of relatively massive stars just after the main sequence stage can be used to estimate the ages of young and moderately old star clusters; these are clusters in which some stars in the mass range we have been discussing have not evolved too far away from the main sequence. From the main sequence lifetimes shown in Table 5 of Chapter 5, this means that the clusters should be no more than a few hundred million years old.

Let us consider first an idealized situation in which we have a cluster of stars of identical chemical composition all of which reach the main sequence at precisely the same time. We are thus considering a group of stars which differ only in mass. As time passes, the stars will evolve away from the main sequence, and because of the dependence of luminosity on mass, the more massive stars will evolve more rapidly. If we plot the evolution of stars of different mass in an HR diagram, we can also insert in the diagram lines of equal time or *isochrones*. Thus, for example, we can mark on each evolutionary track the position of the star after 10^8 years. These points can then be joined up to produce the 10^8 year isochrone. We have already drawn such an isochrone schematically in fig. 30 of Chapter 2 and some more are shown in fig. 69. If all the stars in the cluster had the same age and chemical composition, they should lie on a single isochrone in the HR diagram and the age of the cluster could be deduced.

In practice the problem is rather more difficult. To begin with the theoretical calculations give L_s and T_e while the observations give V and $B-V$. In Chapter 2 we have discussed the relations between L_s and V and T_e and $B-V$ which are needed before a comparison between theory and observation can be made. There are uncertainties in this conversion from theoretical to observed quantities which must not be forgotten. Secondly we cannot expect all stars in a cluster to reach the main sequence at the same time. Stars may not all have been formed at the same time, but even if they were, the time taken to reach the main sequence varies with the mass of the star. If the cluster is old enough, the spread in arrival time at the main sequence will be small compared to the age of the cluster, but for young clusters the isochrones must be drawn after allowance has been made for the time stars take to reach the main sequence. Although cluster HR diagrams are rather sharply

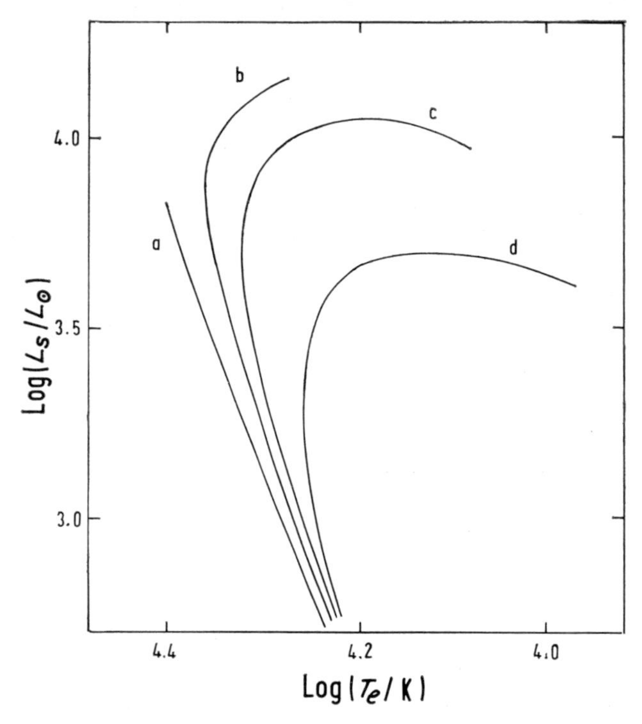

Fig. 69. Isochrones for young clusters. a is the main sequence and curves b, c and d are loci of stars of ages 10^7, $1{\cdot}66 \times 10^7$ and $2{\cdot}65 \times 10^7$ years. The Hertzsprung gap does not appear in this diagram because we have not attempted to show the relative density of stars at different points on the loci.

defined, they are not completely sharp and all the stars certainly do not lie on one theoretical isochrone. However, it is possible to find the isochrone which agrees best with the cluster HR diagram and, using this method, the ages of various galactic clusters have been estimated and the results are shown in Table 9.

It is clear that the simplest idea that all stars in a cluster have exactly the same age and chemical composition is not correct. Although inaccuracies in the observations, variations in chemical composition between the stars in a cluster and the effects of rotation on the properties of some stars all affect this comparison between theory and observation, it seems that the major cause of the spread in the HR diagram of

h and χ Persei	10^7
Pleiades	6×10^7
Praesepe, Hyades	4×10^8
NCG 752	10^9

Table 9. Approximate ages (in years) of young and intermediate age galactic clusters.

young clusters is that there is a finite period of star formation in a cluster. To obtain agreement between theory and observation, it must be assumed that star formation in any one cluster may continue for up to a few tens of millions of years. This is particularly important in the youngest clusters whose age is comparable with this and in some of these star formation is probably still proceeding.

As well as simply comparing the shapes of theoretical and observational HR diagrams, there are other more detailed comparisons between theory and observation which can be made and two of these will be discussed below.

The initial mass function and the relative numbers of red giant and main sequence stars

If we suppose once again that the only important factor differentiating stars in a cluster is mass, we can describe the cluster by what is known as its *initial mass function.* Suppose the number, dN, of stars formed in the cluster with masses between M and $M+dM$ is given by:

$$dN = f(M)dM. \tag{6.8}$$

Then $f(M)$ is called the initial mass function for the cluster. It is called the *initial* mass function because it may be altered as stars evolve if they suffer serious losses of mass. The initial mass function is determined by the processes of star formation. As has been mentioned in Chapter 5, there is as yet no reliable theory of star formation and this means that there is no initial mass function predicted by theory. It is possible to try to obtain some information about the initial mass function from observations and it is of particular interest to try to obtain the initial mass function for different systems of stars to see whether it is always approximately the same. If it were always the same, it would suggest that there was some process which always made a gas cloud fragment in the same way.

It is not easy to obtain the mass function of any group of stars as their masses cannot usually be measured directly and have to be estimated from comparison between theory and observation. E. Salpeter found for the stars in the solar neighbourhood:

$$f(M) = CM^{-2.33}, \tag{6.9}$$

where C is a constant. At present there is no very strong evidence for a different initial mass function in other systems of stars.

When we look at the HR diagram of a cluster, the stars that we see away from the main sequence do not vary very much in mass. This is because the speed of post-main-sequence evolution, once it starts, depends on a relatively high power of the mass of the star. To put it

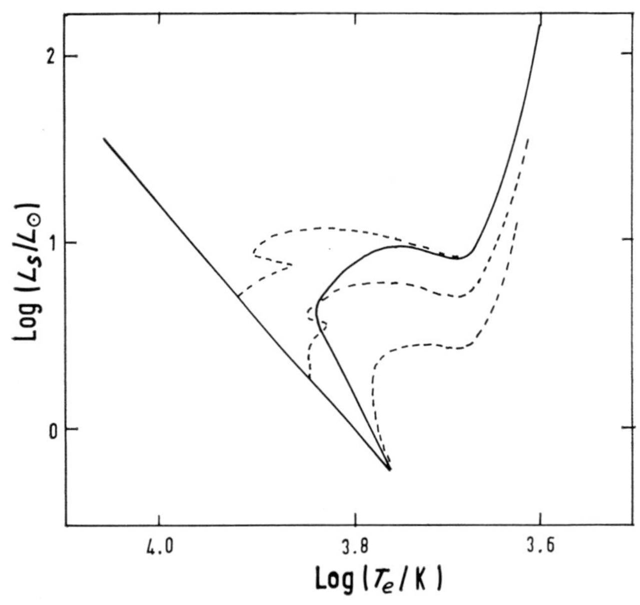

Fig. 70. Isochrones and evolutionary tracks. The three dashed curves are evolutionary tracks for stars of three different masses. The solid curve is an isochrone.

another way, away from the main sequence there is a great similarity between an isochrone and the evolutionary track of a single star. This is illustrated in fig. 70. The evolutionary tracks of stars of different masses are shown and an isochrone is superimposed and this only differs significantly from the evolutionary track of the most massive star near to the main sequence. Provided the initial mass function is not an extremely steeply varying function of the mass of the star, this means that the density of stars in any region of a cluster HR diagram away from the main sequence should be directly proportional to the time that an individual star spends in that region.

We have already used this argument in a weak sense to explain why we do not expect to see many stars in the Hertzsprung gap region of young galactic clusters. We can also compare the number of red giants with the number of stars near to, but above, the main sequence. This is illustrated for the young double cluster of h and χ Persei in fig. 71. Here the observed ratio of stars in regions B and C can be compared with a ratio predicted by theory. At present the agreement between theory and observation is not entirely satisfactory and this may lead to

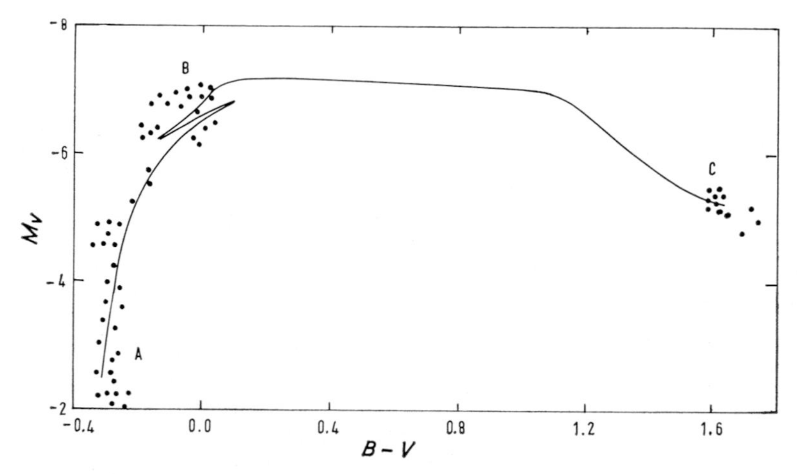

Fig. 71. The HR diagram of the double cluster h and χ Persei on which is superimposed the evolutionary track of a star of $15{\cdot}6 M_{\odot}$.

some modifications in the theory. One problem that affects all comparisons between theory and observation is that stars may suffer instabilities as they evolve, which cause them to lose mass, whereas almost all the calculations made so far have been for stars of constant mass. We shall refer to this possibility again in Chapter 7.

Cepheid variable stars in galactic clusters

We have mentioned several times that there are very few stars in the Hertzsprung gap in galactic clusters. There are a few cepheid variables in galactic clusters. We can see from fig. 28 of Chapter 2 that they lie between the main sequence and the giant branch and they *are* situated in the Hertzsprung gap. As these variables occur in a well-defined region of the HR diagram, it is of importance to ask why stars in that region should be variable while other stars in the cluster are not.

The cepheid variables have a luminosity which varies in a regular, but not smooth manner. Some typical cepheid light curves are shown in fig. 72. The characteristic shape of the curve varies with the period of the oscillation. When cepheid variables, which are named after the star δ Cephei which has been observed since 1784, were first discovered, it was thought that the light variations might be due to eclipses in a binary system, although the shapes of the curves were unusual for binaries. Subsequently study of the spectra of the stars showed that they were undergoing radial pulsations; that is, the radius of the star was varying as well as its light output.

Initially it was thought that the stars were varying because they had received some external disturbance such as might be caused by a close passage of another star. Just as a simple pendulum has a characteristic period of oscillation which is independent of the manner in which it is set oscillating, a star has a characteristic period of radial oscillations. This was calculated and was found to be comparable with the observed

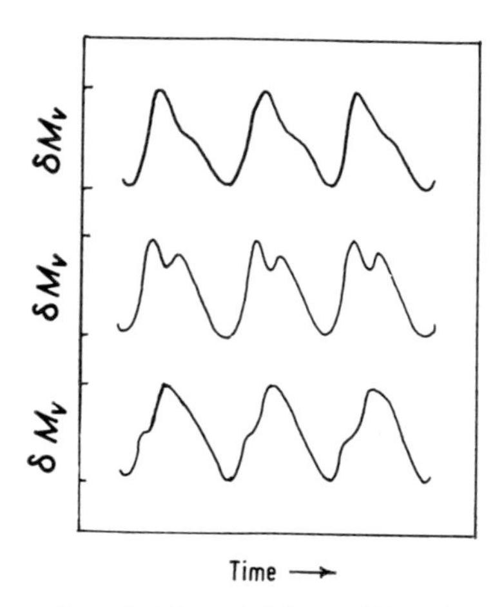

Fig. 72. Light curves of cepheid variables. The characteristic shape of the light curve changes with the period of oscillation. In the figure the top curve corresponds to the cepheid of shortest period.

periods of variable stars. A simple pendulum does not oscillate for ever, as it is gradually damped by air resistance and friction at the support. Similarly internal viscosity and other damping processes affect the oscillations of a star and it was soon realized that oscillations would be damped at such a high rate that it was unlikely that the variable stars arose accidentally. This view was strengthened when it was discovered that the variables occupied a compact region in the HR diagram. It seemed much more likely that the oscillation was stimulated by some process within the star and that this process was operative only for stars in a particular region of the HR diagram.

Recent calculations have given strong support to this view. No stars are in a perfectly steady state: in many cases small departures from the steady state are damped as rapidly as they arise, but in others the fluctuation may be amplified. In the case of stars situated in the region of the Hertzsprung gap, it has been found that small radial oscillations will increase in amplitude and calculations of the growth of these oscillations have gone a long way towards explaining detailed properties of the variable stars such as the shapes of the light and velocity curves. There are still some detailed disagreements between theory and observation, but there seems no doubt that many stars situated in the Hertzsprung gap should be variable stars. The final steady-state oscillation of

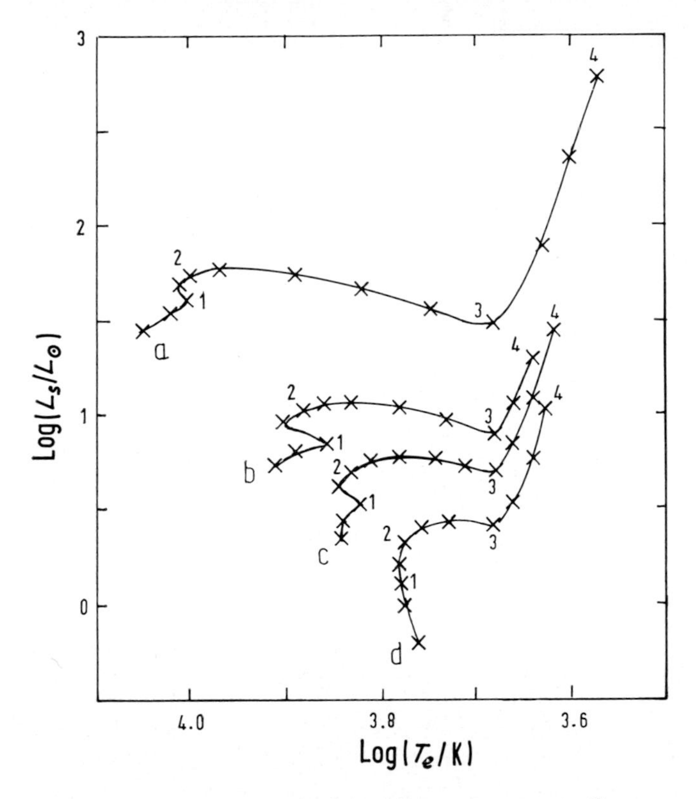

Fig. 73. Post-main-sequence evolution of low mass stars. Curves a, b, c, d are for stars of $2{\cdot}25M_\odot$, $1{\cdot}5M_\odot$, $1{\cdot}25M_\odot$ and $M_\odot$ respectively.

a cepheid variable arises as follows. If the star experiences a small accidental oscillation, this is initially amplified and we say that the star is unstable to small disturbances. As the amplitude of the oscillation grows, the damping forces which we have mentioned earlier also increase and finally a steady oscillation results in which the amplifying and damping processes are just in balance. For the vast majority of main sequence stars the initial small disturbances do not grow and the stars are not variable.

The early evolution of low mass stars

We have mentioned earlier in this chapter two ways in which the evolution of low mass stars differs from that of massive stars. In the first place, when all of the hydrogen has been burnt in the centre of the star, the resulting isothermal core has a mass which is smaller than the Schönberg–Chandrasekhar limit. As a result the star does not move so rapidly to the right in the HR diagram. Secondly, the stars become

$M_s/M_\odot$	1	2	3	4
1·0	6·71	9·20	10·35	10·88
1·25	2·83	3·55	4·21	4·53
1·5	1·57	1·83	2·11	2·26
2·25	0·48	0·52	0·55	0·59

Table 10. Time (in 10^9 years) taken to reach numbered points on evolutionary tracks of fig. 73.

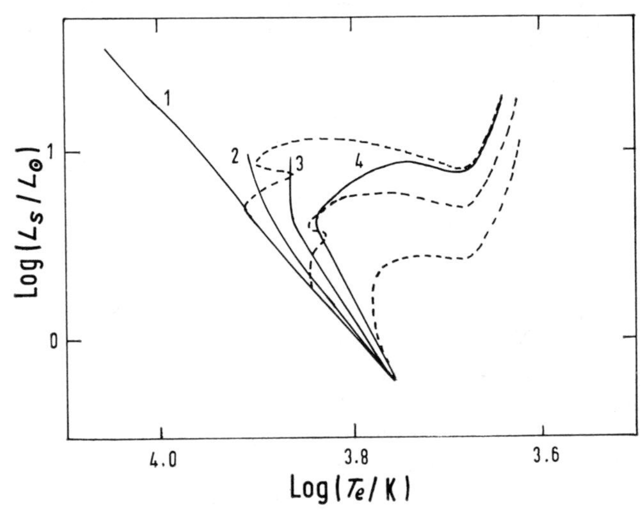

Fig. 74. Isochrones for old clusters. Curve 1 is the main sequence and curves 2, 3 and 4 are loci of stars of ages $0·5 \times 10^9$, $1·5 \times 10^9$ and $2·25 \times 10^9$ years.

degenerate in the centre before the onset of helium burning and because of this what we define to be early evolution ends at the start of helium burning. Recent calculations by Iben of the evolution of low mass stars are shown in fig. 73. As in his studies of relatively massive stars, these calculations are for stars with a population I chemical composition. Thus they are directly relevant to the stars in old galactic clusters rather than the globular clusters whose stars have a much lower abundance of the heavy elements. The times taken to the different points marked on the theoretical curves are shown in Table 10.

Isochrones obtained from the evolutionary tracks of these stars can be used to obtain theoretical HR diagrams for old galactic clusters such as M67 and NGC188. These agree with the observational HR diagrams in having no significant Hertzsprung gap (see fig. 74). As in the case of the young galactic clusters, a cluster age can be estimated by comparison of theoretical and observational HR diagrams and estimated

M67	NGC188	Globular clusters
6	11	15

Table 11. Estimated ages of two old galactic clusters (M67, NGC188) and globular clusters. The ages (in 10^9 years) may be in error by 50% of the values quoted.

ages for these old galactic clusters are shown in Table 11. Other workers have calculated evolutionary tracks for stars in this mass range with much lower heavy element content and these lead to estimates of the ages of globular clusters. These estimates are also shown in Table 11. The accuracy of these ages should not be exaggerated, but it is thought that they are not in error by more than about 50%. This may seem a very large error to be complacent about, but it must be compared not with the small errors of accurate laboratory experiments but with the almost complete ignorance of astronomical ages which preceded this work.

The evolution and age of the Sun

Included amongst the stars of the mass range and chemical composition studied by Iben is the Sun. Although the Sun is in the main sequence region, it must have evolved somewhat from its initial main sequence position. Naturally, as the subject of stellar structure and evolution has developed, a considerable effort has been devoted to trying to account for all of the observable properties of the Sun, about which we know so much more than we know about any other star. Ideally, if we knew the chemical composition of the Sun in fine detail and all of the physical laws accurately, we should calculate its initial main sequence position and then follow its evolution until its actual position is reached and this would then give its age.

In fact there are many difficulties in this procedure. These are not really any worse for the Sun than for any other star except that we can make really accurate measurements of so many properties of the Sun, mass, radius, luminosity, surface temperature, etc., and we expect to find a theory which agrees in detail with all of these. There are un-

shown in fig. 73 but the luminosity reaches a maximum and then declines without helium burning starting. It is not, in fact, believed that stars in this mass range will have completed their main sequence evolution during the lifetime of the Galaxy, because of the dependence of main sequence lifetime on stellar mass.

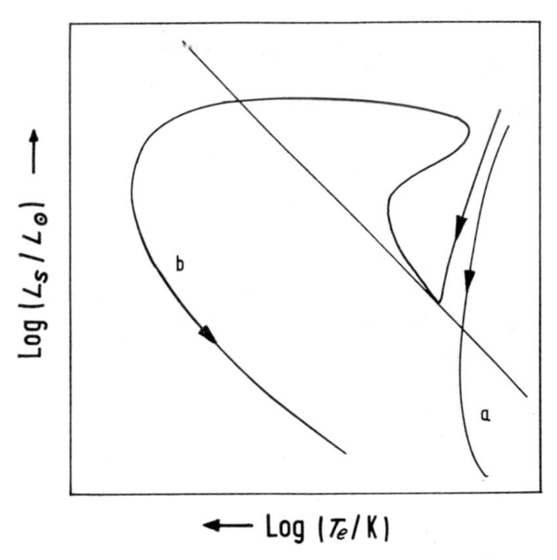

Fig. 75. The evolution of stars of very low mass. a refers to a star of lower mass than $0{\cdot}1M_\odot$ and b refers to a star of mass between $0{\cdot}1M_\odot$ and $0{\cdot}4M_\odot$.

Summary of Chapter 6

The early post-main-sequence evolution of a star depends on its mass. A star remains near the main sequence until most of its central hydrogen has been converted into helium. For high mass stars, with large convective cores, more hydrogen can be burnt before this stage is reached than in low mass stars without convective cores. When the central regions contain no further hydrogen they become almost isothermal. When an isothermal core contains more than about 10% of a star's mass, the central regions contract rapidly and the outside simultaneously expands. This happens almost as soon as the central hydrogen is exhausted in high mass stars and they move rapidly into the red giant region. This accounts for the Hertzsprung gap in some galactic clusters. Low mass stars also become red giants, but less rapidly, and this agrees with the absence of a Hertzsprung gap in globular clusters and the remaining galactic clusters. The calculated evolutionary tracks of stars of different masses, combined with the assumption that stars in a cluster differ mainly in mass, give an estimate of cluster age.

Ages found vary from a few million years for young galactic clusters to more than 10^{10} years for globular clusters.

The straightforward calculation of early stellar evolution ends if the central regions become degenerate while the central temperature is still rising. Nuclear reactions in a degenerate gas may occur explosively, thus resembling a nuclear bomb rather than a nuclear reactor. Because of this, it has proved difficult to study the evolution of low mass stars beyond the start of helium burning. For massive stars, the material remains a perfect gas after helium has burnt. The calculated evolutionary tracks of high mass stars are very complicated with several crossings of the HR diagram. Observations are unlikely to show the full complexity of these tracks because some of the stages of evolution are very rapid and we cannot expect to observe many stars in these stages.

The early evolution of and, particularly, the present state of the Sun has been much studied. A recent attempt to verify conditions in the centre of the Sun by detecting neutrinos emitted in nuclear reactions has given an uncertain result. The number of neutrinos detected has been fewer than expected but the contradiction is not clear.

CHAPTER 7

advanced evolutionary phases

Introduction

In the previous chapter an account has been given of calculations of stellar evolution away from the main sequence. These calculations have not followed a star through its entire life history except approximately in the case of those low mass stars which do not burn their hydrogen and helium before their central temperatures cease to rise and the star as a whole then cools down and eventually ceases to be luminous. For more massive stars there are several evolutionary stages after those which have been discussed in the last chapter. In general, direct calculations of evolution through these phases have not yet been made although some general ideas are available and will be discussed below. The position of some types of star in the evolutionary scheme is still uncertain. These include planetary nebulae, novae and supernovae which will be mentioned further below.

There are several reasons why calculations become more unreliable when we attempt to study the whole evolution of a star. One important difficulty is that in all calculations errors tend to accumulate. The reasons for errors in the calculations are of two types. The numerical processes used in solving differential equations can never be completely accurate and over a long period of integration these mathematical errors tend to pile up. In addition, the mathematical expressions for the physical laws are only approximate. In many cases the physical processes which occur in stars cannot be observed directly in the laboratory and in the case of stellar convection there is neither a good theory nor a good experiment. Small uncertainties in the internal structure of a star may be quite unimportant at an early stage in its evolution, but they might lead to the prediction of an incorrect physical process at a later stage. In the problem discussed in the last chapter, concerning the evolution of stars of five solar masses, whether hydrogen burning restarts in a shell before carbon burning starts in the core cannot be decided except by very careful calculation. It is likely that the deciding factor may involve how much mixing of the stellar material has been produced by convection at earlier stages in the star's evolution.

Two other factors which could complicate the study of late stellar evolution are rotation and magnetic fields. These are relatively unimportant in most stars on the main sequence but, because of the properties of stellar material, they may become more important at later stages of a star's evolution. The viscosity of stellar material is low and its electrical conductivity is high. This means that, as the central regions of a star contract as the star evolves, these regions tend to conserve their angular momentum and also to trap their original magnetic field lines. As a result, both angular velocity and magnetic field strength increase. In a simple geometrical configuration Br^2 and ωr^2 remain constant, where B is magnetic induction and ω is the angular velocity and r the radius of the region considered.

Stellar instability and mass loss

Perhaps the biggest cause of uncertainty is the possibility of instability. In our discussion of stellar evolution we have supposed that a star is spherically symmetrical and that its mass remains constant as it evolves. It is possible that at some stage in its evolution a star might become unstable and lose mass. In order to test this we ought, at each stage in our calculations, to try to discover what would happen to the star if it suffered a small perturbation such as a small compression or expansion in some region or a change in shape. Would the natural tendency be for the compression to increase and for the star to become unstable, or would the compressed region immediately expand to its original state and the star resume its steady evolution? In some cases the physical instabilities may arise without our searching for them. Thus the mathematical inaccuracies we have mentioned above could introduce the small perturbation of the true solution of the equations which is sufficient to trigger off the physical instability. In other cases the equations we are using may not allow the instability to arise and it can only be found by deliberate perturbation of the steady solution.

This is true, for example, if normal stellar evolution is proceeding slowly so that equation (3.4)

$$\frac{\mathrm{d}P}{\mathrm{d}r} = -\frac{GM\rho}{r^2} \tag{3.4}$$

can be used. In most cases, if a small imbalance between the two sides of equation (3.4) is introduced, the star will adjust itself until equality is restored. In other cases the departure from equality will grow and the star will become unstable. Such an instability will only be found automatically if, instead of equation (3.4), equation (3.7) is being used:

$$\rho a = \frac{GM\rho}{r^2} + \frac{\partial P}{\partial r}. \tag{3.7}$$

It might therefore be thought that we should at all times use equation (3.7) rather than the approximate equation (3.4). However, the difference between the two sides of equation (3.4) is normally so very small that serious mathematical inaccuracies are introduced by any attempt to use equation (3.7) and it is much safer to test for stability occasionally than to hope to discover instabilities automatically.

In the last chapter we have mentioned the belief that cepheid variables are stars which are unstable to small disturbances. In their case it does not appear that the instability grows indefinitely, but instead the star settles down in a state of steady oscillation. This is not the only thing that can happen when a star becomes unstable to a small disturbance. Alternatively, the disturbance in the outer layers of a star might grow until the material obtains a sufficiently high velocity to escape from the star. If such mass loss does occur at any stage in a star's evolution, it may alter its whole subsequent life history. Such mass loss in an extreme form certainly occurs in the explosion of supernovae and (to a lesser extent) novae.

The solar wind

We do not know at present how important such loss of mass is. It has been known for some years that even the Sun is losing mass at the rate of between one part in 10^{14} and one part in 10^{13} a year. This loss is known as the *solar wind* because it flows through interplanetary space and past the Earth with a velocity of several hundred kilometres a second. It is too low a rate of mass loss to be observed by direct visual observations, but the solar wind particles have been detected by space probes and the discovery of the solar wind was one of the first astronomical measurements made by the space programme. The solar wind mass loss is unimportant as far as stellar evolution is concerned; it is a direct loss of mass comparable with the mass equivalent of the solar radiation. However, the solar wind has only been detected by careful experiments with space probes and much greater rate of mass loss would remain undetected from more distant stars.

For this reason it is important to try to understand why the Sun is losing mass in the hope that this might give some clues to the probable rate of mass loss from other stars. The Sun has a region just below its surface where theoretical calculations predict that the major energy transport is by convection. This belief is strengthened by the observed appearance of the surface of the Sun. It has a cellular appearance of the type that has been mentioned in connection with convection in Chapter 3 and this is apparently due to the existence of rising and falling elements of matter. It appears that the surface layers of the Sun are *boiling* and that, loosely speaking, they are heating up the outermost layers of the Sun and causing them to be evaporated into space to form the solar wind. If this is correct, we can expect other stars with a deep outer convection zone to lose mass and this seems particularly probable if a star with a deep outer convection zone is also a red giant. The reason for this is that the velocity material has to obtain to escape from a star (or from the Earth for that matter) is $(2GM_s/r_s)^{1/2}$, which for a star of given mass decreases as the radius is increased. We have seen in the last chapter that stars do have deep convection zones when they become red giants. There are some observations suggesting that some red giants are losing mass at a significant rate, but the observations are difficult both to make and to interpret. It is thought that many stars must lose mass as they evolve, but in most cases neither observations nor theoretical studies give a definite answer. One particular problem in which mass loss can produce very interesting results will be mentioned below. This is the evolution of close binary stars.

Having explained the difficulties which occur when one attempts to follow the evolution of a star from birth to death, we shall, in the remainder of this chapter, say something about those groups of stars which are believed to represent later evolutionary stages than the stars whose properties have been studied by direct evolutionary calculations from the main sequence. White dwarfs, which are believed to represent the last phase of stellar evolution, are the subject of Chapter 8.

Globular cluster stars

In fig. 76 we show again the HR diagram of a globular cluster. Direct calculation of the evolution of low mass stars has accounted satisfactorily for the evolution of stars from the main sequence to the onset of helium burning, and isochrones for stars of varying mass have reproduced in reasonable detail the shape of the cluster HR diagram to the top of the giant branch (A). The question then arises; what is the state of stars on the horizontal branch?

In Chapter 6 we discussed how the onset of helium burning in low mass stars is difficult to calculate because the stellar material is degenerate and helium burning may start explosively. If it is explosive, it may lead to the outer regions of the star being blown off into interstellar space. If it is violent, but not quite that violent, the material of the star may be stirred up so that the chemical composition of the star becomes more uniform. When calculations of the helium flash were first attempted, it was assumed that in the initial stages of helium burning the star would move rapidly back down the giant branch to settle on the horizontal branch near B and it would then evolve towards C. This was, I suppose, the simplest suggestion but it did not have any strong theoretical or observational backing and it now appears that stars are more likely to move very rapidly to the left in the HR diagram to join the horizontal branch near to C and mainly evolve to the right.

Whether or not serious mass loss or mixing occurs in the helium flash in low mass stars is still not certain, but its occurrence is suggested below. There is a region on the horizontal branch between C and B which is populated by variable stars of the RR Lyrae type, if these are present in the cluster. Calculations similar to those described for cepheid variables in the last chapter suggest that stars in the region of the HR diagram populated by the RR Lyrae stars should be variable, but this does not explain why some globular clusters contain more than 100 of these variables while other clusters with a similar total number of stars contain virtually no variables. There are certainly small differences in chemical composition between the stars in different clusters and it is possible that these might have a significant effect either on the occurrence of variability or on the speed of evolution through the region where the variables are situated. This is one of those cases mentioned earlier in the chapter where our present knowledge of the physical laws and our present mathematical techniques may be inadequate to give an answer. Theoretical attempts to explain the detailed properties of the variable

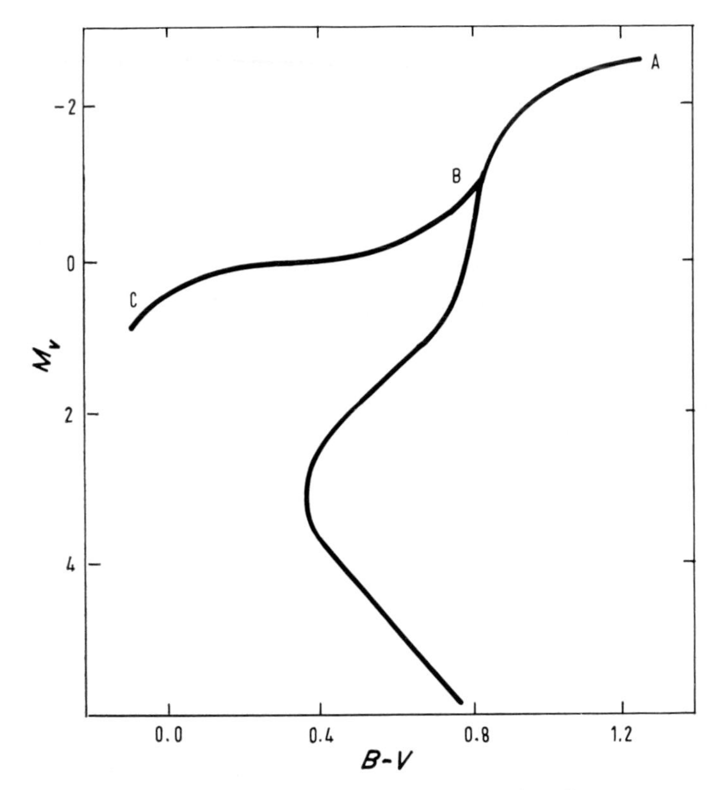

Fig. 76. The HR diagram of a globular cluster.

stars suggest that the variables have a somewhat lower mass than the stars at the top of the giant branch. This could be evidence for mass loss in the helium flash, but at present the difference between the estimated masses of red giants and variables is probably within the errors of the various calculations.

When the stars have moved from C to B along the horizontal branch, calculations predict that they climb up the giant branch again, provided that they are massive enough. At the end of their second ascent of the giant branch, which is probably higher than their first ascent, they are expected to start burning carbon in their central regions. The nuclear reactions concerned have already been listed in equations (6.5)–(6.7).

The stars' interiors are once again very degenerate and there is once again the possibility of an explosive ignition of nuclear fuel which may be accompanied by mass loss. Some people believe that the planetary nebulae, which have been described on page 43, arise at this stage of stellar evolution. They believe that a stellar explosion at the onset of carbon burning produces a roughly spherical expanding gaseous object with a small hot star in the middle. Observations of planetary nebulae indicate that the nebulae *are* expanding and that the central stars *are* contracting. The matter that the star has ejected must gradually become part of the general interstellar medium, while the central stars are believed to continue contracting and to become white dwarfs.

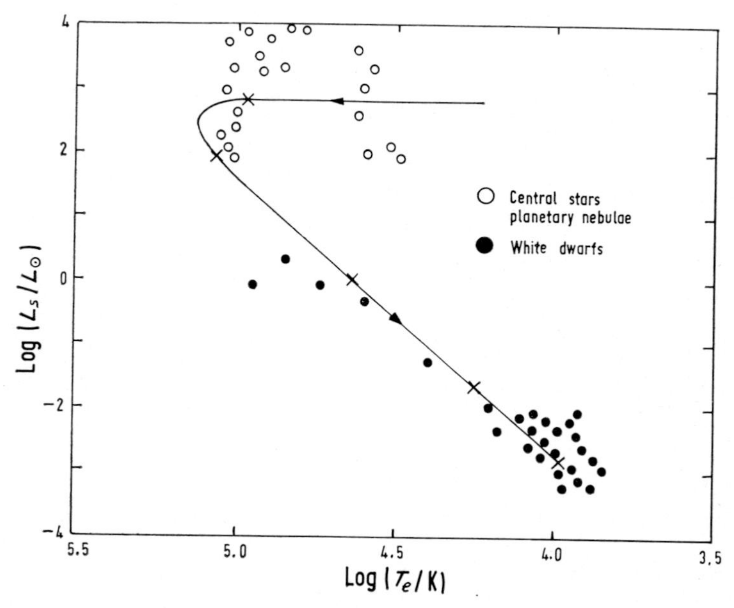

Fig. 77. The HR diagram for nuclei of planetary nebulae and white dwarfs; also shown is the evolutionary track of a star of $0{\cdot}6M_{\odot}$.

The possible connection between planetary nebulae and white dwarfs is shown in fig. 77. An evolutionary track has been calculated for a star of $0{\cdot}6M_{\odot}$, whose only sources of energy are gravitational energy release and cooling. It can be seen that the evolutionary track passes through the regions in the HR diagram occupied by both the central stars of planetary nebulae and white dwarfs. Although $0{\cdot}6M_{\odot}$ is rather lower than the mass usually estimated for stars in active evolutionary stages in globular clusters, it is probably not inconsistent with these if the mass loss accompanying the formation of a planetary nebula is taken into account.

Figure 78 shows a possible complete evolutionary track (post-main-sequence) for a star which is massive enough to burn both helium and carbon, but which never becomes hot enough in its centre for nuclear reactions leading to the most strongly bound nuclei in the neighbourhood of iron in periodic table. It should be stressed that this diagram is only schematic. Different regions of the diagram have been explored, but the evolution of a single star through all of the stages shown has not been studied. The dashed regions of the curve between A and C and D

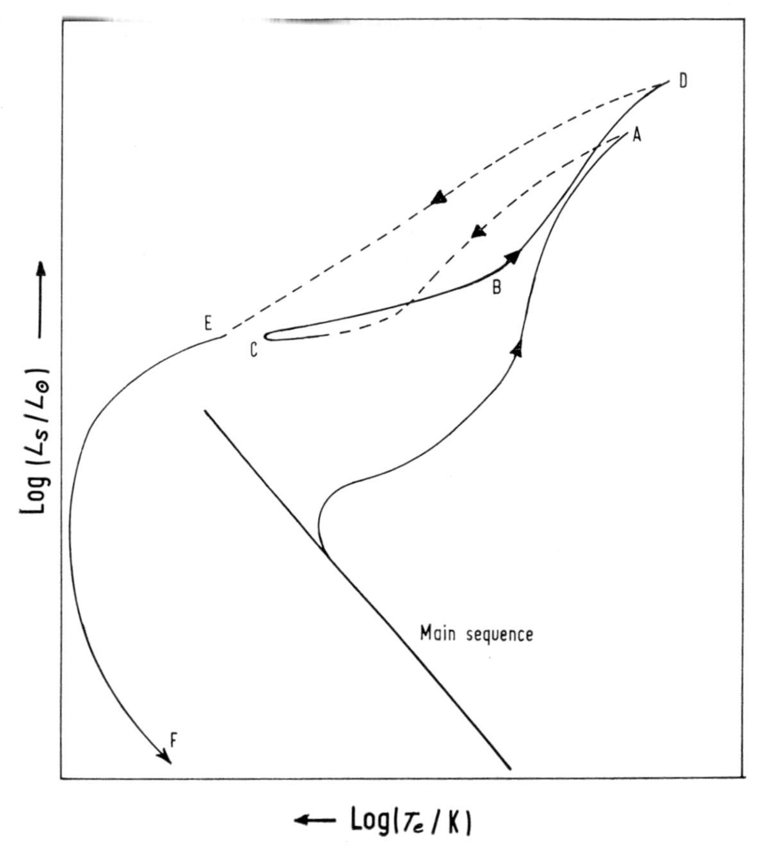

Fig. 78. Schematic evolutionary track of a star of low mass.

and E represent extremely rapid dynamic phases of evolution which have hardly been explored by direct calculations and they are only meant to indicate that we believe that the stars must get from A to C and from D to E somehow. One planetary nebula has been found in a globular cluster and its central star appears to lie a little above the horizontal branch in the HR diagram which is consistent with the above schematic evolutionary track.

It has been suggested that the type of explosion which might produce the planetary nebulae could also be responsible for the occurrence of novae and some supernovae. Supernovae are normally divided into two classes known as Type I and Type II, although there are some supernovae which do not appear to fit into either category. The division into the two types was originally made on the basis of the shape of the supernova light curve. Further study showed that Type I supernovae occurred in regions of galaxies primarily occupied by population II stars, while Type II supernovae were associated with population I stars. It is now believed that Type II supernovae are massive stars and that Type I supernovae are stars of low mass and that the cause of the explosions in the two cases are different.

Both Type I supernovae and novae are objects belonging to population II and they should therefore perhaps arise from globular cluster stars, which are believed to be typical of population II. If supernovae of Type I are essentially similar stars to those which produce planetary nebulae, whether a star becomes a planetary nebula or a supernova may be determined by its precise mass or chemical composition or possibly the extent to which it is rotating. Although the problem is not settled, it appears likely that the nova phenomenon is quite distinct from the supernova or planetary nebula. There is an accumulation of evidence that the novae are all (or almost all) partners in close binary systems and that their instability is associated with this property. Some remarks on the evolution of close binary systems are made later in this chapter.

The late evolution of massive stars and Type II supernovae

We have mentioned several times that when a star becomes degenerate it is possible that its central temperature may reach a maximum value and that the star may subsequently cool down and die (see, for example, page 107), and this is what happens in the case of the globular cluster stars we have described above. The lower the mass of a star the earlier in its evolution this is likely to happen. For sufficiently massive stars, the central regions may not become degenerate until their evolution is essentially over. We can imagine such a star passing through a succession of evolutionary stages in which first one nuclear fuel and then another nuclear fuel supplies the energy which is radiated from the star's surface. As each successive fuel supply is exhausted in the centre of the star, these central regions contract and heat up until the next series of energy-releasing nuclear reactions becomes operative. We have not discussed in Chapter 4 the full range of energy-releasing nuclear reactions which can occur in stars, but we have said that no further energy can be obtained from nuclear fusion reactions once the material has been converted into nuclei in the neighbourhood of iron.

Of course, this will not happen simultaneously throughout the whole star, but initially the centre of the star will reach that state. The variation of chemical composition in a rather highly evolved massive star might appear schematically as shown in fig. 79. When the central regions are mainly composed of iron, they must contract again with no hope that further nuclear reactions will provide the energy which will flow down the temperature gradient which must still exist, if the central regions of the star are to have a pressure gradient capable of balancing the inward gravitational force. Eventually the collapse will bring the central regions to a sufficiently high density that the electrons become degenerate and it might be thought that the contraction would then stop and that the star would cool down like the low mass stars we have described earlier. In fact, in the next chapter, we shall see that there is a

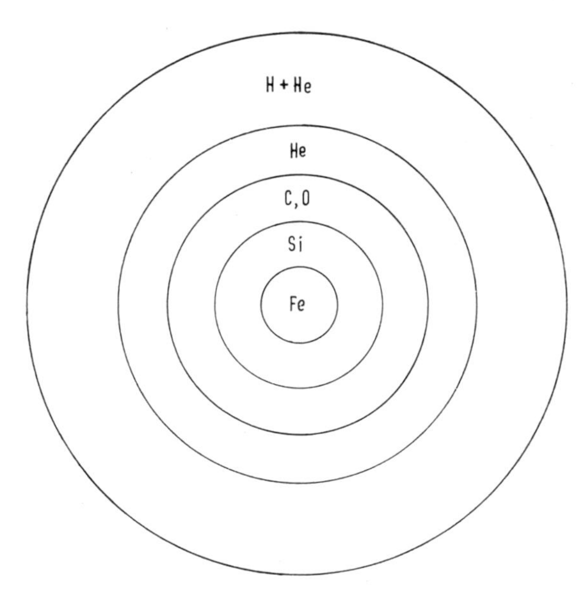

Fig. 79. Chemical composition of a highly evolved massive star.

maximum possible mass for a cold degenerate star. This remarkable result means that even the pressure of degenerate electrons cannot stop a massive star contracting and enable it to cool down and die quietly.

When the central regions of a star are composed of elements such as chromium, manganese, iron, cobalt and nickel, the so-called iron group of elements, does this mean that no further nuclear reactions will occur in the centre of the star? There are no longer any nuclear reactions which release nuclear binding energy in significant amounts. However, if the particles in the stellar material have high kinetic energy, which they do when the temperature is high, there is no reason why nuclear reactions should not occur in which the kinetic energy of the particles is used to enable less strongly bound nuclei to be produced. This is what happens in many artificially induced nuclear reactions in the laboratory when particles, which have been accelerated to very high energies, hit a stationary target. As the temperature at the centre of the star increases, a situation is reached which is analogous to what happens when an atomic gas is heated. Energy can be released by forming atoms out of nuclei and electrons and energy is required to separate electrons and ions, but despite this the gas becomes ionized when its temperature is raised. This ionization is what happens in the early stages of the evolution of a protostar described in Chapter 5. In a similar way, when the material in the centre of a highly evolved star is heated to temperatures of the order of 5 to 7×10^9 K, the nuclei will have a tendency to dissociate, eventually being converted into a mixture of protons and neutrons.

Thus the whole range of nuclear evolution which has been so painstakingly followed in the earlier stages of the star's evolution is essentially reversed. This reversal requires a large supply of energy which initially comes from the kinetic energy of the particles. As this causes their temperature and pressure to fall they are compressed by the outer layers of the star and the central regions of the star collapse very rapidly. A similar rapid collapse occurs in protostars at the time of hydrogen dissociation and ionization, as we have discussed in Chapter 5. This collapse of the central regions means that effectively it is gravitational potential energy which provides the energy to convert the iron peak elements back to protons and neutrons. In the process the central regions of the star may be so compressed that its constituent protons and electrons are forced to combine to form even more neutrons and the star may develop a core of closely packed neutrons. Such a possibility will be discussed further shortly and in Chapter 8.

The theory of Type II supernovae

It should be stressed that the above discussion of the late evolution of massive stars is entirely theoretical. It is, however, believed that the processes discussed above are relevant to properties of Type II supernovae. As we have seen in Chapter 2, supernovae are stars which suddenly increase in luminosity by many orders of magnitude and a supernova at its brightest may give out as much light as a galaxy of normal stars. The supernova outburst is accompanied by a considerable loss of mass from the star and, in fact, it seems likely that a star is shattered by a supernova explosion.

Supernovae clearly cannot radiate for very long at this high rate and they are bright for only a few months and then they become faint and eventually invisible. The light curve of a supernova is shown in fig. 80. At the moment very little is known about pre- and post-supernovae. The last supernova seen in our Galaxy was in 1604 and it is known as Kepler's star because of the careful observations of it made by Kepler. Many supernovae have been observed in galaxies other than our own in the last 30 years and from these observations it has been estimated that, on the average, there might be one supernova per (large) galaxy in about 30 years. This is a somewhat uncertain estimate and it might be thought that it is in serious conflict with the fact that only three supernovae have been observed in our own Galaxy in the past 1000 years. There is not really a discrepancy because there are directions in our Galaxy where interstellar absorption has such a strong effect that not even a supernova would be seen. As we have no real knowledge of pre- and post-supernovae, it is not immediately clear that a supernova is an outburst of a particular type of star rather than an accident that can happen to any star. However, it is necessary that the star concerned be in a position to release a very large amount of energy very rapidly

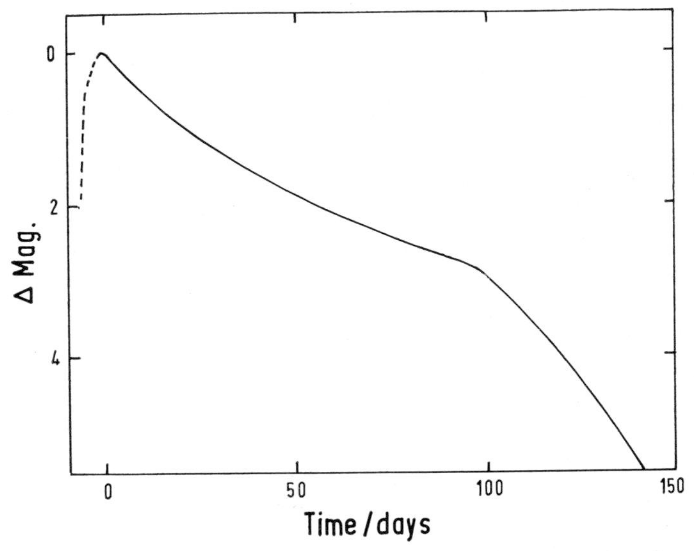

Fig. 80. Light curve of a supernova of Type II. The rise to maximum light shown by the dashed section is usually not observed.

and most ordinary stars cannot do that. Just as it is now thought that variable stars are stars at some particular stage in their evolution, it is probably inevitable that some stars will become supernovae.

We have already made some remarks about the evolutionary state of supernovae of Type I and it is now believed that supernovae of Type II are highly evolved massive stars and that the explosion occurs in the final stages of stellar collapse which has been described earlier in this chapter. We have already seen that the inside of a highly evolved star is likely to collapse, but we now need a reason why the outside should explode. Several reasons have been suggested. They all have in common the requirement that sufficient energy must be deposited or released in the outer regions of the star so rapidly that the outer layers of the star are blown off. As an illustration we will discuss the simplest of the proposed mechanisms. At the same time it must be emphasized that there are other factors which contribute and which could conceivably be more important than those discussed here.

As the central regions of the star collapse, material further out in the star, which still contains potential nuclear fuel, will also fall inwards and will be heated rapidly. Energy-releasing nuclear reactions may then occur explosively in this material and these may suffice to blow off the outer layers of the star and to increase its luminosity substantially. What is left of the star after the supernova explosion may become a white dwarf or a *neutron star*, a star composed almost entirely of neutrons which has a density very much higher even than that of a white dwarf. This will be discussed further in Chapter 8 when we shall learn that, if the remnant is sufficiently massive, present theory does not tell us what will happen to it. In Chapter 8 we shall also have something to say about the relationship between supernovae, white dwarfs, neutron stars and *pulsars*. Pulsars, which are a class of object discovered in 1968, emit pulses of radio waves (and, in at least one case, visible radiation and X-rays) with a pulse repetition rate of the order of a second or less, which is an extremely short time for an astronomical object.

Origin of chemical elements and cosmic rays

There are several reasons why there is a great interest in supernovae other than the ones mentioned above. Many astronomers believe that the original chemical composition of the Galaxy was very simple (possibly it was composed of hydrogen and helium and very little else) and that the heavy elements have all been produced by nuclear reactions in stars. If this is so, the heavy elements which we observe in stars today must have been produced in previous generations of stars and have been subsequently expelled into interstellar space. That explains why there is great interest in the problem of mass loss from stars, which was discussed earlier in this chapter, apart from its effect on stellar evolution. There is the likelihood that many stars lose mass, but there is also the probability that the mass returned to the interstellar medium is just the outer layers of the star which may have the same chemical composition as the star when it was formed. In a supernova explosion there is certainly a large loss of mass from what seems to be a highly evolved star and there is a good probability that the material contains a large amount of heavy elements. In fact, some astronomers believe that most of the heavy elements are produced in stars which become supernovae.

The *cosmic rays* may also be associated with supernovae. Cosmic rays reach the Earth almost equally from all directions. They are very high energy particles travelling with velocities almost equal to that of light and they appear to contain a significantly higher proportion of heavy elements than is found in the atmospheres of stars. It seems possible that in a supernova explosion some small proportion of the material is accelerated to a velocity close to that of light and that this is the origin of the cosmic rays. The suggested association between supernovae and cosmic rays is perhaps natural, not only because of the heavy element content of cosmic rays, but also because a supernova explosion is the most violent event known to occur in our Galaxy. As individual cosmic ray particles have been detected with energies as high as 10^{20} eV, it seems that they must have their origin in a very violent process. At present the association of cosmic rays and supernovae is not unchallenged; some astronomers believe that most of the cosmic rays may have been produced in even more violent explosions such as are observed in some galaxies other than our own.

The Crab Nebula

One of the most remarkable objects known to astronomers is the Crab Nebula. At present it appears to be unique, but this is almost certainly because it is relatively close to us and there are very probably many other objects which are just as remarkable, but which are much more distant. The Crab Nebula is in the site of a supernova seen by the Chinese in the year A.D. 1054† and the nebula is believed to have been produced by the supernova explosion. Optically it is a complex of bright gaseous filaments separated by dark regions. There are several stars in the direction of the nebula but it is not easy to decide whether any of the stars is actually inside the nebula, although as will be mentioned below, it now seems certain that one of them is. The optical filaments are expanding in a way which is consistent with their having been thrown off in a stellar explosion 900 years ago. The Crab Nebula is also a strong source of both radio waves and X-rays; in each case the regions of emission have a complicated structure including what are, with present resolution, point sources of radiation. Recently there has been discovered in the direction of the Crab Nebula a source of radio waves which emits pulses of radiation every 0·03 s. It is naturally supposed that this pulsar, which is in the direction of the Crab Nebula, is inside it and it is felt that it is probably a remnant of the supernova and is also probably a neutron star. This possibility will be mentioned again in the next chapter. The Crab pulsar has now been shown to emit pulses of both optical radiation and X-rays and it has almost certainly been identified with one of the stars observed in the direction of the nebula. The pulsar has now been observed long enough for it to become apparent that its period is lengthening slowly with a characteristic time of about 1000 years for significant change. This ties in with the idea that the pulsar is a remnant of the supernova explosion of 900 years ago. It seems clear that a supernova explosion must be a very complex event and that its consequences are still apparent after 1000

† The Chinese records of novae and supernovae have proved very valuable. They called them guest stars. The value of careful observations to posterity can never have been demonstrated more clearly than in this case.

years. Astronomers would be very excited if there were to be another supernova in our galaxy and one nearer than that in the Crab; though not too near!

Evolution of close binary stars

The structure of close binary stars is complicated because they are near enough for one star to be distorted by the gravitational attraction of the other. Their evolution is even more complicated, but there is one important feature which can be described qualitatively. Consider the evolution of the more massive star of the pair. It will complete its main sequence phase and evolve to the giant branch while the less massive star is still on the main sequence. If the stars are initially close enough together, the outer layers of the newly produced red giant may reach a point where the gravitational attraction of the less massive star will become stronger than that of the star to which the material belongs. In this case the material which passes the zero gravity point will transfer to the other star. There may then be an exchange of material between the two stars to such an extent that the star which was originally the less massive can become the more massive. The exchange of mass only ends when further loss of mass causes the red giant to shrink within the zero gravity surface again.

This type of evolution can have many interesting consequences. It can account for the existence of binary stars in which there is a massive main sequence star with a less massive highly evolved companion. In the past this has been a puzzle because theory always predicts that the massive star evolves more rapidly. Now it seems that the role of the stars has been interchanged after one star has had considerable evolution, but that sufficient additional time has not elapsed for the new massive star to evolve considerably. Sirius and its companion may form such a system. Sirius is a main sequence star with a less massive white dwarf companion.

If the stars are initially sufficiently close together, it seems possible that mass exchange may occur more than once and the originally more massive star might again find itself the dominant partner. Although mass exchange is often expected to occur fairly quietly, it seems that in some circumstances the mass exchange might occur very rapidly and it has been suggested that such catastrophic mass exchange might be the origin of novae. Some post-novae are certainly known to be partners in close binary systems and it is possible that all are, and theories of how such mass exchange between components in a close binary could lead to a nova explosion are being worked out at present.

Summary of Chapter 7

Some stages in stellar evolution have not yet been reached by direct evolutionary calculations from the main sequence. Such direct calculations may be very difficult for several reasons. Errors tend to accumulate and this may make results very uncertain. In some cases, stars become unstable and the resulting periods of mass loss are very difficult to study.

In the case of low mass stars, the first natural break occurs at the onset of helium burning. Provided this process is not catastrophic, the star settles down on the horizontal branch in the HR diagram, but the details of the transition are still not clear. Further evolution takes the star into the giant region again and a second break in direct calculations occurs when carbon burning starts. It has been suggested that the explosive onset of carbon burning leads to formation of planetary nebulae and (possibly) Type I supernovae.

Type II supernovae are believed to arise from massive stars whose central regions have been converted into iron and neighbouring elements. No further energy can be released by nuclear reactions in these central regions but they continue to increase in temperature. Eventually the iron is converted back into helium and neutrons, and the energy required for this conversion is provided by rapid collapse releasing gravitational energy. As a result, the regions further out which still have nuclear fuel are raised to a very high temperature and react explosively to shatter the star in a supernova outburst. Such an explosion may lead to the production of heavy elements and cosmic rays, and what remains may become a pulsar.

Novae are now thought to be partners in close binary systems and a nova outburst may accompany a rapid exchange of mass between the two members of the system.

Because of the way in which they have been obtained, all of the conclusions reached in this chapter are highly uncertain. It would certainly not be surprising if some of them prove to be wrong.

CHAPTER 8

the final stages of stellar evolution: white dwarfs, neutron stars and gravitational collapse

Introduction

IN the previous discussion of stellar evolution it has frequently been remarked that, so long as the stellar material remains in the form of a perfect gas, its central temperature can only increase as it evolves. This result was originally deduced from the Virial Theorem (equation (3.24)) on page 61. As we have mentioned on page 175 in Chapter 7, there is at present no clear solution to the problem of what happens to a star whose central temperature is still rising at the time that nuclear fusion reactions have converted the central regions to iron; in fact, as we shall see in the last section of this chapter, the problem can arise even earlier than that. However, if the centre of the star ceases to be a perfect gas and becomes a degenerate gas, it is possible that the central temperature may pass through a maximum and that the star may cool down and *die*. This possibility has already been illustrated for low mass stars in figs. 75 and 78. Such a dying star is likely to have a low luminosity. It is also likely to have a high density. It can only begin to cool down after its central regions have become degenerate and, if the central temperature has risen sufficiently for one or more sets of energy-releasing nuclear reactions to occur, a very high density is necessary before degeneracy can occur, as has been seen in Chapter 4 (fig. 45).

Such under-luminous dense stars have been observed. They are the white dwarfs which have been described on page 41. All white dwarfs are of low luminosity and those which are partners in binary systems, for which masses are known, are of very high density. Those whose masses cannot be determined almost certainly have a very small radii, as will be discussed on page 187 and it follows that they also have high densities unless their masses are very small indeed. The white dwarfs which have been observed do not have particularly low surface temperatures whereas we might expect dying stars to exist with very low surface temperatures. However, as is shown in fig. 81, the luminosity of known white dwarfs decreases with surface temperature. This means that, even if white dwarfs exist with much lower surface temperatures, they are likely to be too faint to be observed, even if they are quite near to the Sun.

The structure of white dwarfs

Dying stars must be degenerate in their central regions and from fig. 45

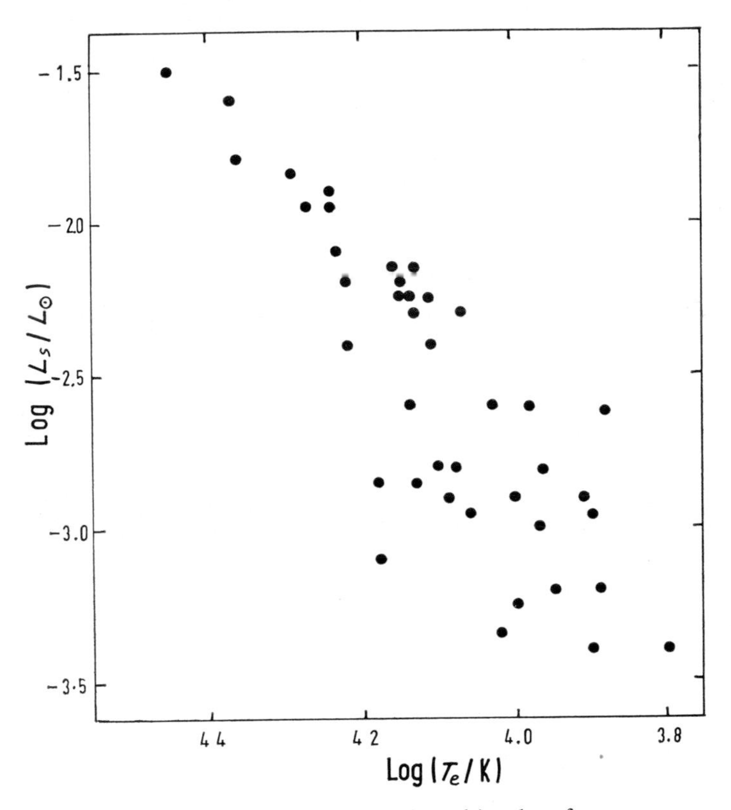

Fig. 81. HR diagram for white dwarfs.

it can be deduced that, as they cool down, they must become increasingly degenerate. Because of this, as a first attempt to study the structure of white dwarfs, we investigate the properties of stars which are made of degenerate gas throughout. In an actual star there is likely to be at least a thin surface layer in which the material is nearer to a perfect gas. In Chapter 4 we have given two formulae for the pressure of a degenerate gas. These are the non-relativistic formula:

$$P_{gas} \simeq K_1 \rho^{5/3}, \tag{4.49}$$

which is valid when the maximum momentum p_0 of the electrons satisfies $p_0 \ll m_e c$ and the relativistic formula:

$$P_{gas} \simeq K_2 \rho^{4/3}, \tag{4.51}$$

which holds when $p_0 \gg m_e c$. We have also stated in Chapter 4 that there must be a gradual change from the first of these formulae to the second as p_0 and ρ increase and the general form of the pressure–density relation can be shown to be:

$$P_{\text{gas}} = f\{(1+X)\rho\}, \tag{8.1}$$

where f is a function which changes from $K_1\rho^{5/3}$ at low values of ρ to $K_2\rho^{4/3}$ at high values of ρ and X is, as usual, the fractional hydrogen content by mass.

The structure of such a fully degenerate star can now be studied by solving the equation of hydrostatic support:

$$\frac{\mathrm{d}P}{\mathrm{d}r} = -\frac{GM\rho}{r^2}, \tag{3.4}$$

and the equation of mass conservation:

$$\frac{\mathrm{d}M}{\mathrm{d}r} = 4\pi r^2\rho, \tag{3.5}$$

in conjunction with equation (8.1.) Because the pressure given by equation (8.1) is independent of the temperature, these equations form a complete set which can be solved without considering the thermal structure of the star. By this, we mean that it is not necessary to study how the temperature varies in the star or how energy is transported. The differential equations (3.4), (3.5) are best considered once again in the form:

$$\frac{\mathrm{d}P}{\mathrm{d}M} = -\frac{GM}{4\pi r^4}, \tag{3.72}$$

$$\frac{\mathrm{d}r}{\mathrm{d}M} = \frac{1}{4\pi r^2\rho}, \tag{3.73}$$

when the boundary conditions to be applied are:

$$\left.\begin{array}{lll} r = 0 & \text{at} & M = 0, \\ \rho = 0 & \text{at} & M = M_{\mathrm{s}}. \end{array}\right\} \tag{8.2}$$

If values of X and M_{s} are specified, the equations can then be solved. Note that in this approximation the chemical composition of the star only enters through the parameter X.

Although the above is formally the simplest way of posing the problem it is not the easiest way to solve it numerically. The reason for this is that the boundary conditions (8.2) are rather awkward as one has to be applied at the centre of the star and the other has to be applied at the surface. There is, in fact, a single infinity of solutions of equations (3.72), (3.73) and (8.1) which satisfy the boundary condition at $M = 0$, each corresponding to a different value of the central density. Only one of these solutions satisfies the surface boundary condition and this must be found by a process of trial and error. It proves simpler to specify the central density of the star, ρ_{c}, instead of M_{s}.

There is only one solution of equations (3.72), (3.73) and (8.1) which has a specified value of ρ_c and $r = 0$ at $M = 0$. This solution can be *followed* outwards until the value of M at which ρ vanishes and this gives the mass and internal structure of the star which has central density ρ_c. This procedure can then be used for a succession of values of ρ_c.

The Chandrasekhar limiting mass

When these integrations are carried out, it is found that the mass is an increasing function of the central density and the radius is a decreasing function of the central density. The more massive degenerate stars are thus smaller than the less massive ones. This is perhaps not surprising when it is realized that the total gravitational attraction which is holding the star together scales as M_s^2, but one result which is at first sight very unexpected is obtained. As higher and higher values of the central density are considered, the mass of the star does not increase without limit, but it tends to a finite limiting value, known as the *Chandrasekhar limiting mass*. For masses greater than this, no models of fully degenerate stars can be constructed. We shall see later that this should be regarded as a mathematical limit rather than a physical limit as the form of the equation of state used, equation (8.1), ceases to be valid before ρ_c becomes infinite. We shall also see that the correction is one of detail rather than principle.

The value of this critical mass depends on the chemical composition of the star through the hydrogen content X. It is possible to show from equations (8.1), (3.72) and (3.73) that there is a very simple dependence on X. The method used is similar to that used in discussing homologous stellar models in Chapter 5. In the present case we simply give the result which can be verified to be correct. If we introduce the quantities $\bar{P}$, $\bar{\rho}$, $\bar{M}$, $\bar{r}$ instead of P, ρ, M and r, where

$$\left.\begin{aligned} \bar{P} &\equiv P, \\ \bar{\rho} &= (1+X)\rho, \\ \bar{M} &= M/(1+X)^2, \\ \text{and}\quad \bar{r} &= r/(1+X), \end{aligned}\right\} \tag{8.3}$$

the equations (8.1), (3.72) and (3.73) become:

$$\left.\begin{aligned} \bar{P} &= f\{\bar{\rho}\}, \\ \frac{d\bar{P}}{d\bar{M}} &= -\frac{G\bar{M}}{4\pi\bar{r}^4} \\ \text{and}\quad \frac{d\bar{r}}{d\bar{M}} &= \frac{1}{4\pi\bar{r}^2\bar{\rho}}. \end{aligned}\right\} \tag{8.4}$$

Equations (8.4) are independent of X and they can be solved to find, amongst other things, the critical value of $\bar{M}$ above which no solution of the equations is possible. When this is done, it is found that the maximum value of $\bar{M}$ is 1·44 $M_\odot$ so that the maximum mass is:

$$M_{\text{crit}} = 1{\cdot}44(1+X)^2 M_\odot. \tag{8.5}$$

This discussion has been for the case of uniform chemical composition. It is more difficult to study non-uniform chemical composition, but a result similar to (8.5) can be obtained with an appropriate mean value of $(1+X)^2$ on the right-hand side.

Equation (8.5) suggests a maximum possible mass of 5·76 $M_\odot$, when $X = 1$, but more detailed study of the problem suggests that this is a totally unrealistic value. A star as massive as 5·76 $M_\odot$ would pass through several of the evolutionary stages described in Chapters 6 and 7 and would burn at least a considerable fraction of its hydrogen into helium and heavier elements. In addition, a comparison of theory and observation for white dwarfs implies that most white dwarfs are unlikely to contain much hydrogen. It appears that the surface temperature of about 10^4 K rises rapidly to a value of order 10^7 K in the interior. At this temperature, if there was any hydrogen present, nuclear fusion reactions would occur and the star would have a much higher luminosity than that observed in white dwarfs. There is, in fact, fairly general agreement that no white dwarfs, other than those of extremely low mass, can contain any hydrogen except in the outermost layers. This implies that the maximum mass predicted by theory is about 1·44 $M_\odot$ (i.e. $X = 0$).

Actual white dwarfs must be somewhat more complicated in their structure than the objects studied here. Their outermost layers will be

$M_s/M_\odot$	0·2	0·4	0·6	0·8	1·0	1·2	1·4	1·44
$\log(r_\odot/r_s)$	1·68	1·81	1·90	1·99	2·10	2·24	2·57	∞

Table 12. Mass–radius relation for fully degenerate stars containing no hydrogen.

a perfect gas rather than a degenerate gas and the thermal properties of the stars must be considered; in the above discussion the temperature and luminosity have not featured. When these effects are included in the theory there is no qualitative change in the results. There is a slight reduction in the predicted maximum mass and other detailed alterations.

For stars of mass less than the critical mass, the solution of equations (8.4) predicts a mass–radius relation which is shown in Table 12. Because we have not calculated the luminosity and effective temperature of these stars, we cannot place them uniquely in the HR diagram. However, for each mass a line of constant radius can be drawn in the diagram and several such lines are shown in fig. 82. It can be seen that these constant radius lines do lie in the region of the HR diagram where the white dwarfs are found, for masses which are comparable with, but not too close to, the critical mass. Individual white dwarfs are thought to follow approximately one of these lines of constant radius as they cool down to their final dead state; there is little contraction of the star as a whole as it finally cools into invisibility.

By saying that the lines in fig. 82 lie in the same region as the observed white dwarfs, we have essentially said that the actual radii of white dwarfs are similar to the radii predicted by the simple theory. In fact, with one exception which will be mentioned below, the radii of white

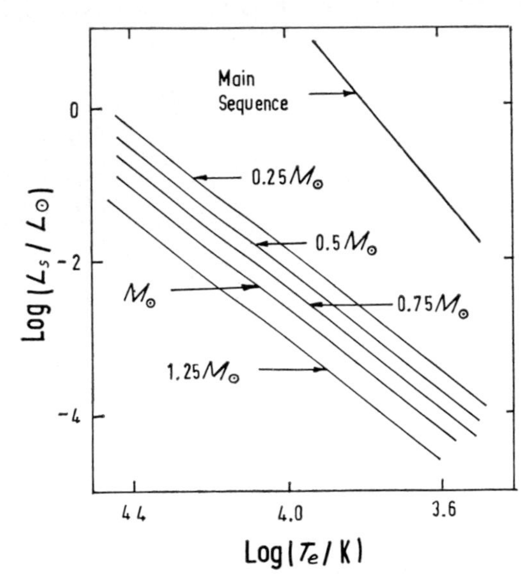

Fig. 82. HR diagram for fully degenerate stars. Such stars have a radius determined by their mass and the constant radius lines for stars of five masses are shown. A cooling white dwarf should approach the appropriate line as its temperature and luminosity decrease.

dwarfs are not really measured directly. From the character of their radiation an estimate can be made of their surface temperature. If it is assumed that this surface temperature is not too different from the effective temperature, an estimate of the radius follows from:

$$r_s = (L_s/\pi ac T_e^4)^{1/2}, \tag{8.6}$$

and this is the method normally used to estimate the radii of white dwarfs. It is these radii which are very close to theoretical values. There seems no doubt that the radii of white dwarfs are very small. The radii of white dwarfs can be measured with difficulty by an observation of the *gravitational red shift* of spectral lines. According to

Einstein's General Theory of Relativity a photon, just like a particle, loses energy in escaping from a region of high gravitational potential. This means that the photon has lower energy when it is received than when it was emitted and this corresponds to a red shift of spectral lines. The red shift in escaping from an ordinary star is very small, but white dwarfs have a much larger value of the critical parameter (GM_s/r_sc^2) and, if their masses are known, radii can be found from the red shift. The values found are in reasonable agreement with those estimated from equation (8.6). As well as discussing the radii of white dwarfs, it is perhaps of more interest to ask what are the masses of observed white dwarfs? Unfortunately only a small number of these are partners in easily observable binary systems, but in all cases in which a mass can be calculated it is less than the Chandrasekhar limiting mass, so that there is certainly no immediate disagreement between theory and observation.

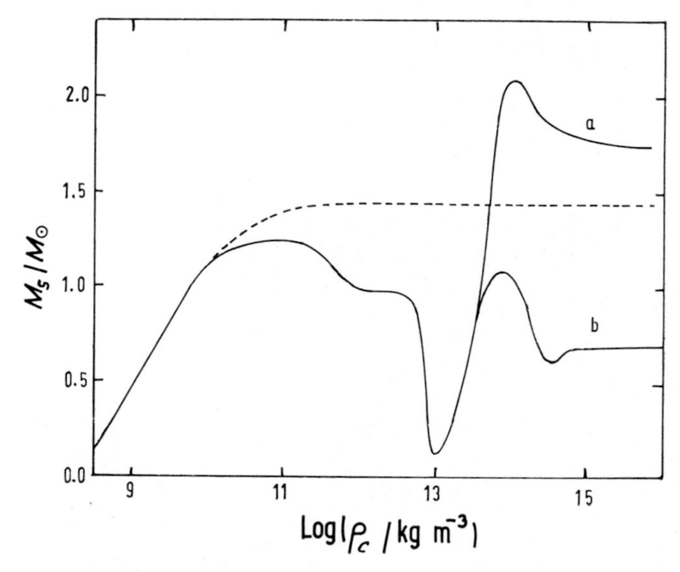

Fig. 83. The relation between mass and central density for white dwarfs and neutron stars. Curves a and b correspond to two different assumptions about the equation of state at high densities. The dashed curve gives the Chandrasekhar relation for white dwarfs.

Neutron stars

We have mentioned earlier that the theory of fully degenerate stars must break down before the central density becomes infinite. In fact, according to theory, when the material is very closely packed with densities in excess of 10^{12} kg m^{-3}, the electrons combine with protons in the nuclei which are present to produce neutron rich nuclei, which would be unstable at lower densities. Eventually at densities in excess of about 10^{15} kg m^{-3} almost all of the material is in the form of neutrons, with just a very small admixture of electrons, protons and heavier nuclei. At these densities the neutrons form a degenerate gas rather than a perfect gas and the star is then known as a neutron star.

If we knew the relationship between pressure and density for degenerate neutrons, which corresponds to equations (4.49) and (4.51) for degenerate electrons, we could repeat the discussion given earlier in this chapter and, for example, obtain a mass–radius relation for neutron stars. Unfortunately we do not have any direct knowledge of the equation of state of material at neutron star densities and there are quite large differences between various theoretical estimates. This means that we cannot present a single unambiguous mass–radius relation for neutron stars. Calculations have been made for several different suggested forms of the equation of state and the results obtained from two of these are shown schematically in fig. 83. Figure 83 shows the relationship between mass and central density for both white dwarfs and neutron stars. All of the suggested equations of state agree that there is a maximum possible mass for a neutron star; unlike the Chandrasekhar theory for white dwarfs, this maximum mass is reached at finite density and at higher central densities the possible mass is lower. What is unclear at present is whether the maximum possible mass for a neutron star is greater than, or less than, the maximum mass for a white dwarf, although the majority of workers in the field believe it is probably of the order of 2 $M_{\odot}$. If this is so, a dying star of mass between $1{\cdot}4M_{\odot}$ and $2M_{\odot}$ is likely to end as a cold neutron star.

Pulsars

Although the possible existence of neutron stars was discussed more than 30 years ago, it is only very recently that any observations have been made which suggest that they do exist. We have mentioned in Chapter 7

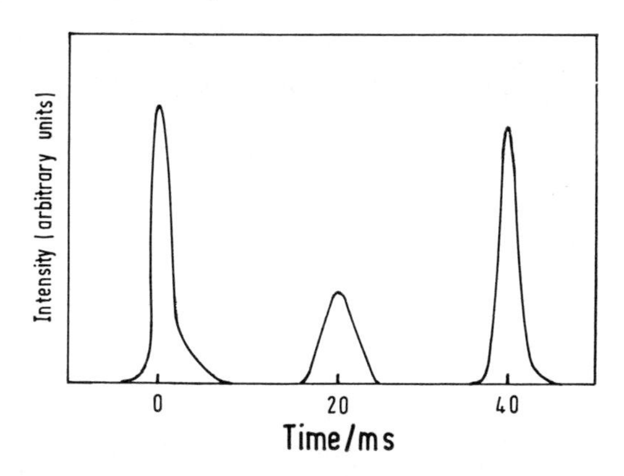

Fig. 84. The radiation from a typical pulsar.

the class of object known as pulsars, which emit periodic bursts of radio waves with a very short period. The radiation from a pulsar may have the character shown in fig. 84; not only is the period short, but the radiation is emitted in two pulses which last only a small fraction of the total period. The character of the pulses shows that the region from which radiation is emitted is very small. If radiation is emitted simultaneously from a region of finite extent, the arrival time of the radiation at an observer varies because of the different distances which radiation from different parts of the source has to travel (see fig. 85). For a pulse to last no longer than about 30 ms, the emitting region should have a linear extent no greater than about 10^7 m and probably much smaller.

If the emission is coming from a reasonable fraction of the object concerned, only white dwarfs and neutron stars seem small enough to be

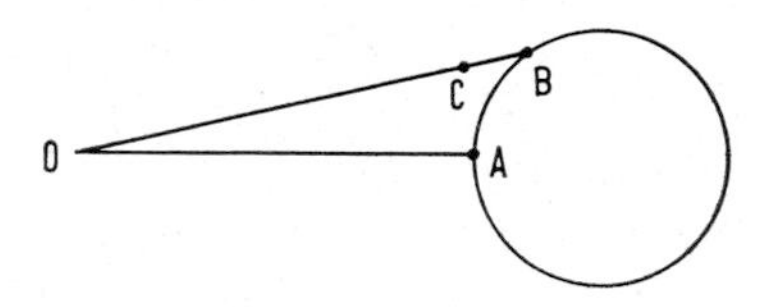

Fig. 85. Light travel time from an extended source. Light from two points A and B takes different times to reach the observer O. Even if a sharp pulse is emitted simultaneously over the source between A and B, it must be spread out on arrival by the time it takes light to travel from B to C.

pulsars. We must also ask what is causing the periodicity and three suggestions have been made. These are that it is a rotation period of a single object, an orbital period of a binary system or a period of radial pulsation, as in cepheid variable stars. Only very small objects can have any of these periods as small as 1 s, if no velocities are to exceed the velocity of light, and now that a pulsar has been discovered with a period as small as 0·03 s it is thought that even a white dwarf is too large and only a neutron star is possible. The subject of pulsars is still young and rapidly changing and the cause of the radio emission is still uncertain. At present the most popular view is that pulsars are rotating neutron stars with radio emission coming from one or two regions near their surfaces.

Gravitational collapse

With the discussion which we have given above of the structure of white dwarfs and neutron stars, it seems that we have an approximate description of the final stage in the life history of stars with masses less than about $2M_{\odot}$. These stars can pass through a sequence of stages of nuclear evolution, their central temperatures can reach a maximum value and then fall and the stars can quietly die. Their life history *may* be more violent and exciting than this, but there is no clear reason why it has to be. However, we now see that we have difficulties in understanding the final stages of evolution not only for stars whose central temperature is still rising, at the time that the central regions are converted into iron, but also for any stars of mass greater than about $2M_{\odot}$. Such stars can apparently only become white dwarfs or neutron stars if they suffer some instability which causes them to eject a considerable fraction of their mass into the interstellar medium.

In the past it has been widely stated that this must be the solution to the problem. Recently there has been an acceptance of the possibility that, within the laws of physics as they are at present understood, there is nothing to prevent a star collapsing to a state of zero radius and infinite density. Such hypothetical objects with densities higher than neutron stars are said to be in a state of *gravitational collapse*. When their densities become extremely high, radiation from them is shifted to the red by the gravitational red shift to such an extent that they become effectively invisible and this suggests the possibility that there could be a considerable amount of *hidden matter* in the universe in the form of gravitationally collapsed objects.

We have stated earlier that the actual value of the maximum mass for neutron stars is uncertain, but that the existence of a maximum mass seems clear. It can be shown that there must be such a maximum mass if the force of gravitation is always attractive, however close together particles are, and if the postulate of the special theory of relativity that no energy can be propagated with a velocity greater than that of light is true. However, the peculiar nature of gravitational collapse has led people to ask whether it is not possible that some of the presently accepted laws of physics go seriously wrong at very high densities.

Summary of Chapter 8

In this chapter we have discussed the final stages of stellar evolution. If the electrons in a star's central regions become degenerate, it is possible that its central temperature may stop increasing and begin to fall. After this, no further significant nuclear fusion reactions occur and the star gradually cools and *dies*. It is believed that white dwarfs are stars of this type. If such a star is sufficiently massive ($\gtrsim 1{\cdot}4M_{\odot}$), the degeneracy pressure of the electrons eventually proves insufficient to resist the attractive gravitational force and the star contracts further. At extremely high densities protons and electrons combine until the star consists almost entirely of neutrons. Such neutron stars are believed to exist and the recently discovered pulsars are thought to be neutron stars. There also appears to be a maximum possible mass for a neutron star. The value of this mass is not well known, but it is thought to be rather greater than the maximum mass for a white dwarf. Above this mass, if the presently accepted laws of physics are correct, objects enter into a rapid gravitational collapse which cannot be halted.

CHAPTER 9

conclusions and probable future developments

In this book we have described the methods used in the theoretical study of stellar structure and evolution and we have discussed many of the results obtained. Our presentation of the subject, with certain limitations to be mentioned below, has been quite up to date. We have tried to discuss the present state of a developing subject and to mention the main uncertainties. As we have stressed, particularly at the end of Chapter 7, some of the detailed theoretical ideas may prove to be wrong, but it is confidently expected that the broad outline of the subject as presented in Chapters 3 to 5 is correct. In this chapter we shall mention briefly some of the limitations of our previous treatment and will describe some of the outstanding problems.

In the first place it is important to realize that, although this book has been written by a theoretical astrophysicist who has a particular interest in obtaining a theoretical understanding of the subject, ultimately all of the theoretical work must be related to observations. This has a two-fold implication. The theoretical worker must keep the observational results in mind and there is a continuing need for new observations. The subject depends considerably on some of the less glamorous parts of observational astronomy. In these days of quasars, pulsars and exploding galaxies, the work of measuring parallaxes and proper motions and studying the orbits of binary star systems is often regarded as very humdrum. However, it is vitally important in supplementing the information which we have about such things as masses, radii and absolute magnitudes. As has been mentioned in Chapter 2, the amount of reliable information which we have about some of these quantities is very small, but further results can be obtained by patient observations.

It should be clear from what has been said earlier in the book that, while no results in the subject can be regarded as exact and not liable to revision, we feel that we have a good understanding of main sequence structure and early post-main-sequence evolution of *single spherical stars*†. There will no doubt be further revisions in the precise values of the opacity and the rate of energy release, and it is to be hoped that eventually a convincing theory of convection will be developed. It is possible that some additional physical processes will be discovered. An

† This restriction means that we are excluding stars which rotate rapidly, have strong magnetic fields or are partners in close binary systems. These will be mentioned below.

example is the following one. A few years ago it was realized that the present theory of the weak nuclear interaction predicts that electrons and positrons can be converted into neutrinos and anti-neutrinos through the reaction:

$$e^- + e^+ \rightarrow \nu + \bar{\nu}. \tag{9.1}$$

This reaction would occur very infrequently in terrestrial conditions and it has not been possible to verify its existence in the laboratory. In highly evolved stars, with central temperatures of 10^9 K or more, electrons and positrons are themselves created from photons by the reaction:

$$\gamma + \gamma \rightarrow e^- + e^+. \tag{9.2}$$

The rate of production of neutrinos by reaction (9.1) is then greatly increased and, as these neutrinos escape from the stars, their rate of energy loss is greater than previously expected. These reactions have been included in some of the calculations reported in Chapter 7. They are believed to be particularly important in massive Type II supernovae.

Refinement of the mathematical expressions for the physical laws will certainly be stimulated by the apparent discrepancy between the number of neutrinos detected from the Sun and the number of high energy neutrinos expected to be produced in nuclear reactions. If this discrepancy is still unresolved after further experiments and a reconsideration of all of the physical laws, this will indicate that there is something radically wrong with our present approach to the theory of stellar structure, but on balance I expect the problem to be resolved within the framework of present ideas. More careful study of main sequence structure and immediately post-main-sequence evolution can be expected to lead to better estimates of the ages of star clusters and of the extent to which correlations exist between stellar ages, chemical composition and spatial distribution in the Galaxy. These latter studies will give information on two important topics which are outside the scope of the present book, the evolution of galaxies and the origin of the chemical elements.

We have seen previously that, even in the case of single spherical stars, there are critical phases beyond which it is difficult to make direct calculations of their evolution. The most obvious case is the start of helium burning in low mass stars. There is a variety of problems involved in extending the study past the existing difficult points. One is the sheer size of the computers which would be required. At any stage in a stellar evolution calculation it is necessary to have a complete description of the entire structure of the star stored in the computer. Clearly, the number of points in the star, for which values of all the physical quantities must be stored, is greater for stars with a complicated internal structure. This is particularly true when a star has several zones in which different nuclear reactions are occurring and when some parts of the star are expanding and others are contracting. If very detailed mathematical expressions are used for the laws of opacity and energy generation, a large space may be required for the storage of opacity and energy generation tables or for the expressions from which these quantities can be calculated. These problems concerned with storage of information are being overcome as larger computers become available. Secondly, when a star is evolving very rapidly, such as when there is an explosive release of nuclear energy, the interval of time which can be allowed between successive models of the star may become very small, if an accurate result is to be obtained. In this case the amount of computing time required to study significant evolution may become prohibitive. This again is becoming less troublesome with the advent of larger and faster computers. Finally there are problems which arise if a star shows unpredictable behaviour. In some cases the usual form of the equations of stellar structure may not be capable of describing the behaviour. For example, when a star loses mass there may be difficulties in formulating an entirely satisfactory surface boundary condition as it may not be clear where the surface of the star is; the lost mass must at some stage cease to be part of the star, but when exactly?

It is certainly to be expected that further progress will be made in the study of the evolution of single spherical stars. Evolution will be studied through further periods of nuclear energy release in the case of relatively massive stars. For low mass stars, more elaborate mathematical techniques may enable evolution to be followed through the onset of helium and carbon burning, along the dashed sections in fig. 78 of Chapter 7. There is also likely to be a better understanding of the way in which a star becomes a variable star when it enters a particular region of the HR diagram and to be more reliable predictions of when mass loss is likely to occur. I will, however, hazard a guess that it will be a long time before there is a direct evolutionary calculation through and out of a phase of mass loss.

Very little has been said in this book about stars other than those that are single and spherical. The reason for this omission is three-fold. In the first place, the majority of stars are effectively single and spherical; secondly, most of the work on non-spherical stars is very recent and general conclusions are still unclear; and thirdly (and more important), a discussion of the theory of non-spherical stars requires mathematical techniques more advanced than those that are appropriate to this book. It is, however, perhaps worth while to conclude the book by mentioning some of the problems of interest.

Rapidly rotating stars and stars containing strong magnetic fields are both distorted into a spheroidal state. If a star is both rapidly rotating and contains a strong magnetic field and the axes of rotation and magnetism do not coincide, the star may be even less symmetrical. In many (but not all) ways the effects of rotation and magnetic fields are similar and we will discuss only rotation. There are two types of effect produced by rotation. The first is the structural effect that stars are non-spherical. This leads to a flow of energy which is also not spherically symmetrical and to a variation of effective temperature over the surface of the star. This means that the apparent luminosity of a star depends on the inclination of its axis of rotation to the line of sight. It is also found that the total luminosity of a star is altered by rapid rotation. Because of these effects it is important to discover how much of the observed spread in main sequence luminosity at a given colour index can be accounted for by effects due to rotation. This is particularly important in several young clusters, such as the Pleiades, which contain many rapidly rotating stars.

Rotation also has some more subtle effects on the thermal properties of stars. We have already mentioned these briefly in Chapter 6 where we have said that it used to be thought that meridional circulation produced by rotation kept stars well mixed. Although it is now thought that such mixing is unimportant in most stars, it should be significant in some rapidly rotating stars. As well as mixing the stellar material, such circulation has an effect on the rotation law of the star. The viscosity of stellar material is not very effective in reducing relative velocities between elements of stellar material in the star's lifetime. This means that each element tends to retain its own angular momentum about the star's axis of rotation as the law of conservation of angular momentum applies to each element individually. As meridional circulation moves an element around, it moves into regions previously occupied by elements with a different amount of angular momentum and this modifies the distribution of angular momentum and angular velocity through the star. Recently there has been considerable interest in trying to discover whether there are any special laws of rotation which are not altered by meridional circulation. If these exist, we might expect a rotating star ultimately to settle down with such a distribution of angular momentum. The best that can really be said at the moment is that the problem is extremely complex and the results are not entirely clear, but it seems possible that there are no steady rotation laws.

The study of close binary stars is in a more primitive state than the study of rapidly rotating stars. We have mentioned in Chapter 7 that mass exchange between the components is likely to play an important role in the evolution of close binary stars and recently many calculations have been made of ways in which such mass exchange might radically influence the progress of stellar evolution. In these calculations the distortion in the shape of a star by its close companion has been estimated only very crudely. The problem of calculating the exact shape of members of a close binary system is very difficult. The whole system must in any case rotate and it does not possess the spheroidal symmetry of a single rotating star.

There is one problem of particular interest connected with the pre-main-sequence evolution of close binary stars. If stars are very close (and some are, in fact, in contact) when they are main sequence stars, what were they like at an earlier stage in their life history? Assuming that both components of the binary system have contracted towards the main sequence from a state of low density, there appear to be two general possibilities. The first is that the system was originally one star and that some instability caused the star to fragment not long before it reached the main sequence. This fission theory of the origin of binary stars was the first theory suggested. At one stage it was thought that it had been proved mathematically that such fission was impossible, but the status of this proof is now less certain. The second suggestion is that as the stars contract some process coupling the spin angular momentum of the individual stars to their orbital angular momentum causes them to move closer together. Reasons for both of these ideas have been suggested, but it is not yet entirely clear which one is correct.

This brings us back to the part of the subject which is, perhaps, most in need of progress. Although we have stressed several times that main sequence structure and post-main-sequence stellar evolution can be studied without there being a good theory of star formation, the subject would look more tidy if such a theory existed. The reason for difficulties in studying star formation is not hard to find. A typical density in the interstellar medium is 10^{-21} kg m^{-3} while a typical stellar density is 10^3 kg m^{-3}. The temperature of the interstellar gas is often less than 100 K compared to temperatures inside stars of order 10^6 K. The study of star formation thus requires the treatment of an extremely wide range of physical conditions. In addition, simple considerations suggest that objects of star cluster size first form in the interstellar medium and that individual stars are then formed by fragmentation. The study of such a fragmentation process is bound to be difficult.

The book should not, however, end on a pessimistic note. It is true that there are many unsolved problems concerning stellar structure and evolution, but the amount of progress that has been made in the subject is very great. Observations of the surface properties of stars have been combined with a few basic physical laws and many measured and calculated properties of atoms and nuclei to give us what we believe to be reasonably accurate information about the unobservable interiors of stars. Unfortunately these stellar interiors are likely to remain invisible for ever. For that reason particular interest will be paid in the next few years to the ultimate resolution of the present discrepancy between theory and observation in the solar neutrino experiment. If, as seems quite likely, theory and observation can be reconciled, the theory of stellar structure will appear to have particularly good foundations.

APPENDIX

thermal equilibrium

If a physical system is isolated and left alone for a sufficiently long time, it settles down into what is known as a state of *thermal equilibrium*. In thermal equilibrium the overall properties of the system do not vary from point to point and do not change with time. Individual particles of the system are in motion and do have changing properties. For example, electrons may be being removed from and attached to atoms. There is, however, a *statistically steady state* in which any process and its reverse occur equally frequently. Thus, in the example mentioned above, the number of atoms ionized per unit time is equal to the number of recombinations. Because the properties of a system do not vary from point to point when it has reached thermal equilibrium, all parts of it have the same temperature.

If two such isolated systems are brought into contact, heat will flow from one to the other until they reach the same state of thermal equilibrium and hence the same temperature. In thermal equilibrium all of the physical properties of the system (such as pressure, internal energy, specific heat) can be calculated in terms of its density, temperature and chemical composition alone. In nature a true state of thermal equilibrium will be approached closely but never quite reached.

As mentioned above, in thermal equilibrium the temperature of the system is the same everywhere and it does not change with time. Inside a star, the flow of energy from the centre to the surface implies that there must be temperature differences and these are, in fact, very large. However, provided that the temperature change between two successive points at which a particle of the system suffers a collision and points of emission and absorption of a photon is small compared to the temperature itself, there can be a very close approach to thermal equilibrium. These conditions are very easily satisfied in the interiors of stars and, as a result, we are able to assume that quantities such as pressure, opacity and rate of energy generation are functions of temperature, density and chemical composition alone.

In thermal equilibrium the intensity of radiation is given by the Planck function $B_\nu(T)$. An interesting feature of this radiation intensity is that it does not depend on what matter is present, although the time that it takes to reach its equilibrium intensity does depend on the matter. In most experiments on Earth, conditions are far from true thermodynamic equilibrium, because the amount of radiation present is far below its thermal equilibrium value. Inside stars, in contrast, the radiation intensity is very close to the value predicted by Planck's law and the radiation energy density and radiation pressure may be important.

INDEX

SUGGESTIONS FOR FURTHER READING

It is not easy to find other books at, or near, the level of this book. The following are suggested:

A. J. Meadows, *Stellar Evolution* (Pergamon Press). This is a semi-popular account with no detailed mathematics.

F. Hoyle, *Frontiers of Astronomy* (Heinemann, Mercury (paperback)). This semi-popular book covers a much wider range of topics than stellar evolution. Many of its details are out of date, but it gives a very lively picture of modern developments in astronomy.

Two rather more advanced books, of which the second gives an observer's view of the subject, are:

M. Schwarzschild, *Structure and Evolution of the Stars* (Princeton, Dover (paperback)).

W. Baade, *Evolution of Stars and Galaxies* (Harvard).

A book in the Wykeham series on a related subject is:

H. R. Hulme and A. McB. Collieu, *Nuclear Fusion.*

THE WYKEHAM SCIENCE SERIES

for schools and universities

1 *Elementary Science of Metals* — J. W. Martin and R. A. Hull
(S.B. No. 85109 010 9) — **20s.—£1.00 net** *in U.K. only*

2 *Neutron Physics* — G. E. Bacon and G. R. Noakes
(S.B. No. 85109 020 6) — **20s.—£1.00 net** *in U.K. only*

3 *Essentials of Meteorology* — D. H. McIntosh, A. S. Thom and V. T. Saunders
(S.B. No. 85109 040 0) — **20s.—£1.00 net** *in U.K. only*

4 *Nuclear Fusion* — H. R. Hulme and A. McB. Collieu
(S.B. No. 85109 050 8) — **20s.—£1.00 net** *in U.K. only*

5 *Water Waves* — N. F. Barber and G. Ghey
(S.B. No. 85109 060 5) — **20s.—£1.00 net** *in U.K. only*

6 *Gravity and the Earth* — A. H. Cook and V. T. Saunders
(S.B. No. 85109 070 2) — **20s.—£1.00 net** *in U.K. only*

7 *Relativity and High Energy Physics* — W. G. V. Rosser and R. K. McCulloch
(S.B. No. 85109 080 X) — **20s.—£1.00 net** *in U.K. only*

8 *The Method of Science* — R. Harré and D. Eastwood
(ISBN 0 85109 090 7) — **25s.—£1.25 net** *in U.K. only*

9 *Introduction to Polymer Science* — L. R. G. Treloar and W. F. Archenhold
(ISBN 0 85109 100 8) — **30s.—£1.50 net** *in U.K. only*

10 *The Stars: their structure and evolution* — R. J. Tayler and A. S. Everest
(ISBN 0 85109 110 5) — **30s.—£1.50 net** *in U.K. only*

11 *Superconductivity* — A. W. D. Taylor and G. R. Noakes
(ISBN 0 85109 120 2) — **25s.—£1.25 net** *in U.K. only*

THE WYKEHAM TECHNOLOGICAL SERIES

for universities and institutes of technology

1 *Frequency Conversion* — J. Thomson, W. E. Turk and M. Beesley
(S.B. No. 85109 030 3)

2 *The Art and Science of Electrical Measuring Instruments* — E. Handscombe
(ISBN 0 85109 130 X)

Price per book for the Technological Series **25s.—£1.25 net** *in U.K. only*

NOTES